THE
Next Great
Thing

ALSO BY MARK L. SHELTON

Working in a Very Small Place

THE
Next Great
Thing

The Sun,
The Stirling Engine,
and
The Drive to Change the World

MARK L. SHELTON

W · W · *Norton & Company* · *New York* · *London*

FIRST EDITION

The text of this book is composed in Galliard with the display type set in Caslon 471. Composition and manufacturing by the Maple Vail Book Manufacturing Group. Book design by Marjorie J. Flock.

This work was funded in part by a fellowship from the Ohio Arts Council

Library of Congress Cataloging-in-Publication Data

Shelton, Mark L.
    The next great thing: the sun, the Stirling Engine, and the drive
to change the world / Mark L. Shelton.
        p.   cm.
    Includes index.
    1. Solar engines.   2. Stirling engines.   3. Sunpower (Firm)
I. Title.
TJ812.5.S54   1994
338.7'62142—dc20   93-34409

ISBN 0-393-03619-7

W. W. Norton & Company, Inc., 500 Fifth Avenue, New York, N.Y. 10110
W. W. Norton & Company Ltd., 10 Coptic Street, London WC1A 1PU

1 2 3 4 5 6 7 8 9 0

Dedicated to the memory of
Ross Shelton (1907–1993),
practical engineer

There is nothing in the history of technology in the past century and a half to suggest that infallible methods of invention have been discovered or are, in fact, discoverable. It may be true that in these days the search for new ideas and techniques is pursued with more system, greater energy and, although this is more doubtful, greater economy, than formerly. Yet chance still remains an important factor in invention and the intuition, will and obstinacy of individuals spurred on by the desire for knowledge, renown, or personal gain the great driving forces in technical progress. As with most other human activities, the monotony and sheer physical labor in research can be relieved by the use of expensive equipment and tasks can thereby be attempted which would otherwise be wholly impossible. But it does not appear that new mysteries will only be solved and new applications of natural forces made possible by ever increasing expenditure. In many fields of knowledge, discovery is still a matter of scouting about on the surface of things where imagination and acute observation, supported only by simple technical aids, are likely to bring rich rewards.

—from *The Sources of Invention,* by John Jewkes, David Sawers, and Richard Stillerman, 2d ed. (New York: W. W. Norton, 1969), p. 169.

Contents

Acknowledgments

Mark Shelton is here to investigate the possibility of writing
a book about Sunpower.

—*Monday morning meeting minutes*

WHEN I FIRST SHOWED UP at Sunpower, I'm sure, no one had any idea what to expect. That sentence from the meeting minutes of my first day in the building is remarkable for its lack of certainty; several people told me later that they assumed I would drop the subject like a hot potato as soon as I found out what Sunpower was like or that, like so many things at Sunpower, it was an intriguing idea, but one that would never "make it into hardware." That this book has made it into hardware is a testimony less to my perseverance than to the unique cooperation I was given by everyone at this unusual little company.

William T. Beale invited me into his world and by his example encouraged others to do the same. To have someone let you dog his steps and watch over his shoulder for almost two years is more than any writer can reasonably hope for, but William not only let me; he often tracked me down when something particularly interesting was going on. I can't thank him enough.

Elaine Mather, Peggy Shank, and Dolly Walker started doing me favors small and large on my first day and continued to do so until my last. Neill Lane, who originally piqued my interest in solar-powered Stirling engines, spent more hours than I am sure he cares to remember explaining to me everything from the Stirling cycle to the "pi and pi-squared" theory of how long projects take; as will become evident, there would be no book without the contributions of Neill. The other engineers on the solar project were equally patient and helpful, and became friends as well as mentors and instructors: Barbara Fleck, Abhijit Karandikar, and Joerg Seume deserve special thanks.

Jarlath McEntee and Gong Chen are not only fine engineers but fine human beings: open, patient, thorough, goodwilled; Jarlath is a natural teacher, and a good one. David Berchowitz, John Crawford, and Robi Unger were under especially intense pressure during the time I spent at Sunpower, but this didn't stop them from spending time and energy with me; Nick van der Walt and Chris Capers made the small-engine test cell an oasis of good information and good cheer. Eric Bakeman and Morgan Mitchell know, I am convinced, everything there is to know about computers; Scott McDonald and Robert Redlich, everything about electricity; Mark Labinov, about higher mathematics; Lyn Bowman, about electronics and billiards. Only rarely did their countenances betray the fact that I had asked an incredibly simple (and, occasionally, silly) question, which they then proceeded to answer.

The job title "Test Engineer" is a special honorific at Sunpower: Al Schubert and Dave Shade were the men who, finally, made so many of the engines go. Their knowledge is encyclopedic and their skills legendary, and rightfully so; Dave Shade even managed to teach me something of machining, as well as many other things. The Sunpower technicians had the often difficult task of putting together the designs and ideas of the engineering staff, frequently contributing ideas and designs of their own: Larry Haas, Jim Smailes, Trevis Thompson, Paul Moore, Pat Dailey, Bob Black, and Charles Stoneburner were very skilled, and very helpful to me, Dale Kiikka, Chuck Howenstine, Ben Wilkus, and Hans Zwahlen in particular. Mike McCune, Janet Halbirt, Ken Kelly, and their easygoing supervisor, Todd Cale, explained not only what they did, which was draw Sunpower's machines, but how they did it. In Sunpower's famous machine shop, Jerry Royse, Ray Klinebriel, Doc Licht, Mike Stoneburner, Sr., and Mike Stoneburner, Jr., deserve mention and much thanks, as do Sunpower's unofficial corporate historian, Andy Weisgerber, Laurie Zucker, Debbie Harris, Neil Barron, Debbie Barron, Barbara Withem, Rita Withem, the late Larry Nelson, Mohammad Dehghani, Bjorn Carlsson, Mary Fabien, Israel Urieli, Meyer Esmaili, Gordon Brunson, Gary Wood, Jeff Schrieber, John Bean, and Rocky Kubo. John Woodrow had several frank and helpful discussions with me. Sean Ramsay helped me in the early stages with research.

I used many fine reference works, several of which deserve mention and thanks. David Berchowitz's dissertation at the University of the Witwatersrand has been extremely useful as a source of information on the practical dynamics of Stirling engines, which is not to say I understand all that is contained in it—those who do form an exclusive club indeed. I am pleased, too, to acknowledge and thank P. W. Atkins of Lincoln College, Oxford, for his fine and accessible book *The Second Law*, an excellent introduction to the history of mechanical engineering in the context of thermodynamics. Several of my examples of the second law have their genesis in this lovely book. My first resort, as always, has been to *Van Nostrand's Scientific Encyclopedia*, a gift to me from my mother, and no better gift to the scientifically curious exists. From my father, I received two books that stood him well in a half century of practical engineering: *Audel's Answers on Practical Engineering* and *Audel's Answers on Refrigeration*.

I would be remiss if I did not mention Tim Seldes and Miriam Altshuler of Russell & Volkening, my much admired and skilled literary agents, and Starling Lawrence, my editor at W. W. Norton; both he and Donald Lamm, Norton's president, supported this project when it was merely the twinkling of an idea. Every writer should be so lucky.

Finally, I again thank my wife, Eve, who was helpful, supportive, patient, and deeply interested throughout my time at Sunpower, as she always is.

Athens, Ohio
January 1994

Author's Note

I HAVE TRIED TO BE as faithful as possible to the words and language of the people at Sunpower, who spoke frankly with me at every turn, and thus, I believe, deserve to be quoted frankly and fully in recompense. There are, however, two exceptions.

The first is mainly a question of style. Sunpower is a multicultural and multinational place where the language in use, while not exactly polyglot or a lingua franca, to some degree retains the native accents of the speakers: the South Africans have South African accents (which to most Americans, including me, sound vaguely British), the Russians Russian accents, and the native southeastern Ohioans an accent of their own: Jarlath McEntee, who is Irish, *sounds* Irish; Abhijit Karandikar, who was born in London to Indian parents and raised in Canada and California, speaks flawless American English with only the slightest accent, but the accent is Canadian; and so on.

I have not tried to reflect these many and varied accents in any way, except in one or two cases where they are important to the context. The alternative—to try to replicate the sounds of the voices by means of devices such as the apostrophe and the contraction, or to attempt to invent words and spellings that resemble accented sounds—would be, in my view, unnecessarily distracting. Instead, I have chosen to render voices in a virtually accentless prose. The words, however, are as accurately theirs as I could capture them in my notebooks.

The second exception is also intended to avoid unnecessary distraction, but of a different sort. What are referred to as "good old Anglo-Saxon words" show up in the speech of most people with varying degrees of regularity. One word in particular is used by the people of Sunpower as a sort of all-purpose modifier, intensifier, clarifier, and, only occasionally, expletive, by some people often, by others less so, and by a few rarely or never. Some reserved it for the most extreme situations; others used it quite amiably in response to, say, a question about lunch.

I rarely heard one person use this word in anger directed at another. I did, however, hear it used countless times in reference to machinery (lathes, drill presses, telephones, fax machines, computers), weather, measuring instruments, writing implements, delivery schedules, windows, doors, metal, plastic, wood, glue, electronic equipment, government, government agencies (one in particular might as well have the initial *F* precede the three more well-known initials on its official seal), thermodynamic phenomena, accidents, fate, and, perhaps most often, Stirling engines.

In all but the rarest occurrences, the word was used as a sort of linguistic marker or spacer, or in a way that seemed unconscious. Sometimes it was used in ways that were, in context, highly entertaining; other times, it was used in ways that suggested only a verbal habit. (I should mention that I picked up this verbal habit myself, which, once acquired, is a hard one to break.) To include this word every time it was used would possibly illustrate just how unconscious its use seemed to be, but it would also, I expect, become irritatingly repetitive. Besides, it would lengthen the book by many hundreds of words. The alternatives—letter and dashes, euphemisms, ellipses—are unattractive in both appearance and fidelity, so I have excised it from the text. To the charge that this is a form of bowdlerization, I plead guilty with an explanation: my goal is not to clean up the English language, or to polish the image of Sunpower and its employees, or to seem to be passing some sort of judgment. Instead, I have tried to illuminate and explain the work that goes on at Sunpower as accurately as possible, and it is a mark of the access and cooperation granted me that I sometimes heard words that people might not say in front of their mothers. I hope my portrayal is accurate and real; if it is not, however, the reason is not that I've refrained from including this good old Anglo-Saxon word in the text.

M. L. S.

THE
Next Great
Thing

Prologue:
Two Parties

I

BOOOOOOZE OR NO BOOOOOOZE?" calls out Peggy
Shank, her fists full of dollar bills (left hand for
"booze"—that is, beer; right hand for "no booze"—
flavored seltzer waters), as she bangs through the big blue swinging
doors into the machine shop at Sunpower, Inc. It is a Friday afternoon
in December, and while the technical milestone for a party (generally
defined as a first-time run of a newly designed Stirling engine) hasn't
been . . . *technically* met, it has been quite a while since the last party, so
an encouraging and important test of the design of the biggest, newest
engine in the place, the five kilowatt, has been determined by the party
committee (which consists of Peggy Shank) to qualify. Outside, it is
cloudy and warmish, not the sort of weather usually associated with
Ohio, but fairly typical for a December afternoon in Athens, which is a
small city closer to Kentucky (to the south) and West Virginia (to the
east) than to, say, Cleveland, which lies four hours north.

In the halls and warrens and laboratories and cell-like rooms called
test cells that fill the ramshackle brick building on Athens's West Side,
Sunpower employees each in turn dig into the pockets of jeans and into
purses, and shout back at Peggy either "booze" or "no booze," and
ante up.

"What's the occasion?" asks Jerry Royse, although he is already
reaching for his wallet.

"We had a successful . . . something or other on the 5K," says
Peggy, hands out. "Booze or no booze?"

It is getting on dusk when the beer and seltzer arrive, along with bags of potato chips and pretzel sticks, and roughly half of Sunpower, Inc., crowds into the company kitchen and reaches for bottles, either booze or no booze, and handfuls of chips. The noise level starts out low, but builds, resonating through the tiny, Spartan room, with its ancient refrigerator, its table covered in oilcloth, its half dozen mismatched chairs, and its stack of large orange plastic recycling tubs labeled "aluminum," "glass," "paper," "mixed paper," and "Stirling engines." (This last bin is, at the moment, empty.) As the volume of the voices grows, the focus grows more acute; by five o'clock, it seems everyone in the room is talking about—what else?—Stirling engines.

There is, at this moment, plenty to talk about. Sunpower, which was at its birth, on June 1, 1970, a company of one—William T. Beale, the founder, president, and technical director—had, on this December afternoon, more employees (forty-eight), more ongoing projects (seven), and more corporate and government sponsors for those projects than it had ever had in its history. It also had more potentially *commercial* products being made into hardware than ever before. It is necessary to say "being made into hardware," because William Beale has an inventory—partly on paper, but mostly in his head—of certainly hundreds, and on some days, it seems, thousands, of potential applications for Sunpower Stirling engines: it is hard to name a process into which William Beale has not imagined insinuating a Stirling engine.

Even the list of projects making their way into hardware is impressive for a tiny (by corporate standards) research and development company with virtually no capital, no manufacturing expertise, and no out-and-out successes in its history: a small Stirling engine analogous to a remote generator (with funding provided by a "private sponsor"—that is, a person or corporation that is paying for the development of a prototype); a heat pump aimed at home heating and cooling (with R & D money from a division of the federal Department of Energy); a refrigerator (two sources of funding: one that has just run out was seed money from a large American corporation; just starting is a small grant from DOE for thorough testing and evaluation of the prototype); a smaller, more specialized cooler (a private sponsor); a solar-powered engine suitable for providing home and industrial electricity, intended to hook into the established power grid (another private sponsor, the largest manufacturer of diesel engines in the world, a company looking to diversify). The company has money, and work, and good employees, and good progress. On party afternoons, the future of Sunpower seems as limitless as the power of the star it was named for.

This party attracts most of the usual suspects. Peggy Shank, whose official title is something like "Administrative Assistant/Publications," but whose job description would read more like "William's Right Hand," is there, as is Elaine Mather (job title "Administrative Manager"; job description, "Run the Place."), as are Andy and Dolly and one of the Debbies and Neil and Neill and David and Dave and Ben and Trevis. William himself is there, sipping a beer and giving the lie to at least two "truths" around Sunpower—that he never goes to the parties and that he never has a beer. William even tells a joke (another Sunpower "truth" to the contrary, that William has little sense of humor), which is all it takes to get Ben Wilkus started on a long stream-of-consciousness round of jokes, asides, funny stories, and imitations of everyone at Sunpower. Jarlath McEntee, who is as Irish as he sounds, shows up in a leather motorcycle jacket, a T-shirt, and jeans; he spent yesterday dressed in an immaculate white shirt and a stylish tie on account of visitors from the outside world, in this case from the DOE's Oak Ridge National Laboratory, who oversee the heat pump project of which Jarlath and Gong Chen are the engineers and designers of record. So today it is as if he dressed down with a vengeance.

Nick van der Walt and Chris Capers come in from the small-engine test cell; although Nick is from Zimbabwe and Chris from Appalachia (two of the more different places on the planet), they work together so well, and so thoroughly, that their names are said as one: "Nickandchris." On the blueprints from which the machine shop makes parts for the various engines, there is a small box for the initials of the designer of each part—the piston, say, or the cylinder, or the alternator, or the gas spring, each engine being at least the sum of its many parts (or, as one of the mantras at Sunpower has it, "Parts is parts"). Some of the drawings for the small engine have "NV" in the box and some have "CC," but quite a few have either "NC" or "CV," which almost no one notices, but which the drafting department (which inserts the initials) finds highly amusing. The small engine is such a collaborative project that it is not surprising that Nick and Chris show up for the party at the same moment.

At about this point, and after telling one more joke, William leaves, knowing that in some ways he is a damper on the parties (at one Monday morning meeting, he makes light of this, saying, "For party scheduling purposes, be aware that I'll be out of town from Wednesday to Monday this week"). If he is a damper, it is not necessarily because of his sense of humor, although the jokes he finds funniest tend to have something to do with either engineering or thermodynamics, neither of which is generally considered to be a rich topic for humor (his last joke of the day is the old chestnut of a man falling through the air grappling with a faulty parachute, who meets another man flying *up* through the air; the skydiver asks him if he knows anything about parachutes; the man on his way into orbit asks if the parachutist knows anything about Coleman stoves); rather, it is because William is so intertwined with everything that happens, for good or ill, at Sunpower, that when Sunpower people talk about Sunpower, they are, finally, talking about William Beale.

In one corner, Lyn Bowman is quizzing Neill Lane about Neill's engine, which is known officially as "the solar five kilowatt" and colloquially as the "5K." Lyn is an electrical engineer with a background in various arcana of microelectronics, and thus used to contemplating machinery that is shaped like a postage stamp and that would be lost inside a thimble; the 5K has the shape of a battering ram and the size of a large motorcycle. In its pressure vessel, the engine looks something like a bomb, a resemblance that has been pointed out countless times to Neill Lane and that he has disputed every time. "It's *not* a bomb," he says; "it's an engine, and watt for watt, it's *smaller* than any engine Sunpower has ever made. You want a lot of kilowatts, you need lots of engine."

The 5K is in many ways an immensely technical and complicated project, bigger than anything Sunpower has done in hardware before; thus it, even more than the small engine (whose power output is aimed for 200 to 250 watts, or a fifth to a quarter of *one* kilowatt), is a collaborative project. But it is also generally thought of as "Neill's engine"; he wouldn't be blamed for feeling somewhat proprietary, although he rarely acts proprietary. Nonetheless, he is the 5K's strongest defender against skeptics.

Five kilowatts of alternating current would power a dozen reasonably energy-efficient households. And because the 5K will make electricity at the same voltage and phase that traditional power plants do, it can be hooked into the existing "grid"—into the thousands of miles of wires and thousands of transformers, meters, substations, and electrical outlets already established in the outside world—the world, that is, outside of Sunpower.

The company employees work with a vengeance (changing the outside world can be hard), but on this day they don't exactly party that way. Many of their appearances, like William's, are almost pro forma: show up, have a beer, tease someone mercilessly, eat some organic potato chips, and head out. What they head out *to* is a world that seems replete with possibilities, with expansive dreams. Given that their company now has more money than it has ever had, and more hardware, and more talent, they can be forgiven for thinking not that the party is over but that it is just getting started.

II

That party took place in December, which can often be warmish and gray in Athens, Ohio; spring tends to be hottish and gray and humid. For all the difference that a few months can make, the weather can be irritatingly similar. That anyone would have settled in southeastern Ohio in the first place is a mystery to Neill Lane, who comes from South Africa. "If *I* would have come through here, I would have just kept on *going*," he often says. "It's the most inhospitable climate I've ever *seen*." Neill is saying this while sitting at the bar of the West End Tavern, at what has been billed as a party, but which really seems more like a wake: Sunpower, Inc., which three short months ago was at its biggest size, has suffered its most violent contraction in history. No one got paid the previous Friday, and no one got paid this day; that reality, while unpleasant, is not unprecedented in the history of small, highly speculative R & D companies working on unproven technology. For that matter, the layoff of almost half the company is not unprecedented either, although when a company of twelve lays off half the staff (as Sunpower did five years earlier) it affects fewer people than when a company of fifty does so.

Sunpower has traditionally repaired to the West End for celebrations, so it is fitting that Sunpower should go there for a postmortem. The bad news, rumored for weeks (first there was a moratorium on salary for senior employees; then there was a pay cut, even for those on salary moratorium; having your pay cut when you aren't even getting paid is perhaps the best fertilizer for rumor and speculation on a company's future), has hit home. Some engineers are gone, or will be going shortly; so are draftspersons, accounting and professional support staff, machinists, technicians. A wide swath has been cut through the company, and the West End, a bar famously no-frills in a town of no-frills bars, once again hosts the men and women of Sunpower for a farewell bash. At the bar, Neill Lane quits complaining about the weather and asks Jarlath McEntee if he wants to buy a bicycle, only half jokingly: Neill's and Jarlath's working hours will be cut approximately in half for the foreseeable future, an economy that comes on top of two weeks without pay. Neill also savors a line he is itching to use on William, when next they argue about the 5K engine—"You can't *fire* me, because you don't *pay* me"—a line, however, he will not get the chance to use.

Eric Bakeman strolls through the door of the West End in tennis shorts and a T-shirt. *"Norm!"* a half dozen people shout at his arrival: Eric has been known to suggest repairing to the West End for virtually any reason, and he resembles in demeanor that genial man-on-a-bar-stool from the television show "Cheers." Although Eric will remain as a "consultant" to the company, since he wrote much of the computer code that Sunpower uses to simulate and test Stirling engines, he too has been laid off.

Next to Neill Lane sits David Berchowitz, fellow South African, fellow graduate of the University of Witwatersrand, fellow senior design engineer, and *notorious* Sunpower optimist; David is trying to put the best face on things. "When the book about Sunpower is finally written . . . ," he begins, but Neill finishes the sentence for him, to great laughter, "it will be a *pamphlet!*"

"It'll stop in the middle of a sentence!" suggests Andy Weisgerber.

"They won't have the money to print it!" offers Jarlath.

"No, no, no, the publisher will go broke *before* they print it!" says Ben Wilkus.

"The writer will be laid off."

"They'll repossess the typewriter!"

"No, no," David tries again, "I'm serious; when the last chapter on Sunpower is written . . . ," but again, the interruptions are raucous.

"It'll be a best-seller on the fiction list."

"No, the *science* fiction list."

"No, the science fiction and *fantasy* list."

"No, no, no. No one at Sunpower will be able to read it, 'cause they can't afford to *buy* it."

David Berchowitz tries one last time. "Listen. At the end of the day, the *technology* will have the last word; all of this piddling about money will be a *footnote.* The financial part of it is *nothing.* The *technology* will be the story."

"David," Neill Lane says with finality, "when the book about Sunpower is written, it's going to end with . . . *Chapter Eleven!*"

Everyone at the bar cheers.

Chapter I

The Magic Box

IN ATHENS, OHIO, the city-supplied recycling bins are a bright pumpkin orange, as if to make it easier to see just who *is* recycling and who is not. So every trash day looks like the trash day after Halloween in Athens, a small and poorish town in southeastern Ohio that tries to take its earthly responsibilities seriously: there is a flash of orange at the end of virtually every driveway.

Athens is this kind of place: it features a worker-owned bakery and a worker-owned restaurant, a shop that sells low-slung, three-wheeled commuter bicycles ("Peach Ridge Pedal Power"), a non-smoking coffeehouse, a non-smoking health food restaurant, and a peace vigil every weekday at noon outside the courthouse at the corner of Court and Washington streets ("Won't you stand five minutes for PEACE?"), holidays included.

Athens is the home of the Appalachian Peace and Justice Network, the Appalachian Ohio Public Interest Campaign, the Appalachian Mother Earth Network, the Worker Owned Network, the Coalition for Ohio Appalachian Development. There is a thriving chapter of the Society for Creative Anachronism, men and women in medieval robes who play darts each Monday night in O'Hooley's bar. The *Mother Earth News* once did a fancy, full-color spread on Athens, citing much of the above in support of its contention that for those who love Mother, Athens is quite the place to live. Athens has a tie-dyed clothing store, a bead store, a half dozen "intentional communities" (to use the contemporary term for what were once called communes), and a shop that sells and services fiddles and guitars, with a bench out in front for strumming. There is a state university, Ohio's oldest, with some seventeen thousand students, most of whom arrive in Athens from the suburbs of Ohio's large cities and *gawk*, having for the first time in their short lives seen a real live hippie. The students arrive via U.S. 33 or U.S. 50, Athens having no interstate, and although the students outnumber the year-round residents, it's easy sometimes for them to feel outnumbered, to feel like outsiders, especially if they arrive on trash day. If they arrive on trash day, they see, at house after house after house, bright orange plastic recycling bins that sport the logo "Athens Recycles." It reads like an admonition.

A first-time visitor to Athens would easily find the Ohio University campus, resplendent in red brick and white trim and towering oak and sycamore trees. It dominates the center of town. Many of the other Athens landmarks are also prominently located, well known, and easy to find. By contrast, it's likely that a first-time visitor would find Byard Street only by accident, if at all. And even once found, Byard Street is unassuming in the extreme, easily forgotten. There are houses typical of the Athens West Side (small frame houses on tiny lots), a bank of utilitarian apartments, a railroad siding, and a red brick warehouse or factory of some kind, a long stretch of building that shows its age.

Hardest of all to see, unless one is looking for it, is a small steel plate with a mysterious, almost cabalistic symbol, attached to a rickety Cyclone fence surrounding the long, low brick building on a hidden side street in Athens. At first glance, it looks like Aladdin's lamp. The symbol is actually a stylized fulcrum in black, with an orange ball suspended over one end. It represents, according to its creator, "an unbalanced environment, with the sun—the orange ball—on one side; as with a scale or a mathematical equation, something is needed to balance the power of the sun and bring the system and the symbol into equilibrium." Beneath the fulcrum, unbalanced on one side by the sun, are the words "Sunpower, Inc. Stirling Engines." Both sets of words, those on the plaque and those describing its significance, belong to a man named William Beale.

Although the words appear beneath the Aladdin's lamp of a logo, they would more properly fit above the fulcrum, balancing the sun and bringing the symbol into equilibrium, because the goal of the company is to harness the power of the sun and, in doing so, to bring the world— the *world*—into equilibrium. Sunpower, Inc., is a company created for, and dedicated to, principles that are lofty even when measured against the standards of a town with a daily peace vigil and an aggressive recycling program, principles that incorporate nothing less than a fundamental and total change in the way the world makes and uses energy. The people of Sunpower are not out to make the world a little better; they are not out to make a contribution on a scale analogous to the little gush of virtue that comes from hauling a bright orange plastic tub full of aluminum cans out to the curb each Wednesday. They are out to *save the world,* and to do so by the development of the Stirling engine.

To say that the people of Sunpower, Inc., are out to save the world is problematic, though, because it is so easy to say. The phrase "to save the world" passes by in a moment, vanishes into the places where hyperbole must disappear before the serious business can begin. So it bears repeating, amplification: *Sunpower, Inc., is out to change the way the world makes and uses energy.* In a world where the soft targets, like the recycling of aluminum cans, are not only easy to hit but immediately rewarding, Sunpower has chosen the hardest target of all, a target all but impossible to see, take seriously, or even imagine. Converting the world to Stirling-engine power is not one of the fifty simple things that people can do to save the earth. Instead, it is one of the fifty hardest. It provides no gush of warmth, no immediately visible result, no smile of self-satisfaction. The slogans of environmentalism fit on T-shirts and bumper stickers. Sunpower is more like an unabridged copy of *War and Peace.*

What the people of Sunpower want to do is this: they want to take two of the most ingrained infrastructures in the industrialized world and convert them to an alternative technology that is as potentially revolutionary as it is unproven. Instead of meeting the world's energy needs by burning fossil fuels, damming rivers, and splitting atoms, Sunpower intends to develop and foster the implementation of a pollution-free substitute: instead of electricity produced by the combustion of coal, the world will use electricity produced by the thermal conversion of the sun's heat into usable power. Or, as William Beale likes to put it, "on an area 170 miles by 170 miles in the Arizona desert—a *postage stamp*—the sun can provide *all* of the energy consumed in the United States."

That's one goal. The company also wants to replace the technology by which the world cools itself with an alternative device that uses no chlorofluorocarbons, which, as even small children these days know, are damaging the earth's protective ozone layer.

A Stirling engine, simply, is an external-combustion engine, with a piston that moves inside a cylinder; it differs from most other engines because it may use virtually any fuel as a heat source, rather than being dependent on a fuel such as gasoline, which is expensive and environmentally harsh. It is easy to describe a Stirling engine this way; it is even easy to make a simple Stirling engine to demonstrate the principle. When William Beale demonstrates Stirling technology, he uses a Sunpower engine called the B-10, which is about the size of a large flashlight; it has a cylinder, a piston, a displacer, and a round steel "heater head," to which William applies heat with a propane torch. When he heats up the head for a few moments and then raps the engine on a tabletop, the piston bobs up and down. Making a B-10 is simple. It is a terrific visual aid. It is when one tries to apply the technology to large-scale energy projects like electricity generation and refrigeration that the whole idea gets astonishingly complex.

William Beale did not find either of these tasks—a changeover from fossil fuels to sun power, a changeover from chlorofluorocarbons to environmentally gentle refrigeration—in a book of easy things to do to save the earth. He has not recruited a team of engineers to accomplish these tasks from around the country and around the world by offering big money, extensive perquisites, pastoral working conditions, or the promise of quick reward. The people of Sunpower work, on the average, harder, longer, and with fewer resources than they would at virtually any other engineering or technical job in the country, and they do so not only without the promise of success but without even the promise that the company they work for will last long enough to see through even the preliminary tasks they must accomplish before they can get to the really hard parts. If there is a heaven, there is most probably a special corner reserved for the people of Sunpower, but in this world they must be satisfied with small rewards indeed. They work in uncharted territory, in the terra incognita of invention and engineering and design, and they are inventing every step of the way. What they are inventing are technologies that, in order to be considered even remotely successful, must be absolutely invisible to the end user: if the world is to run on solar-powered Stirling engines, then solar-powered Stirling engines must tie in seamlessly to the existing electrical infrastructure that ends in every electrical socket in every dwelling in the world.

Even their intermediate tasks, which are modest in the face of a revolution so comprehensive, seem large indeed: they have not yet made *one* full-size solar-powered Stirling engine capable of tying into the existing electrical grid, let alone the ten thousand machines necessary for a full-scale demonstration of the technology, let alone machines enough to cover an area 170 miles square in the Arizona desert, which is what it would take to power the United States with Stirling engines, something like 100 million of them.

The scale of the Sunpower dream is at once exhilarating and disturbing. Exhilarating, because the task is so large. Disturbing, for the same reason. The very existence of Sunpower, Inc., makes the recycling of an aluminum can seem a tiny and hollow act, because a world that can be saved by the recycling of an aluminum can seems manageable, benign, comforting. A world that has to be saved by millions of machines in the Arizona desert is anything but. If the people of Sunpower are wrong, they are crazily wrong. If they are right, if their dream is necessary to save the earth, then the consciousness of most Americans needs to be expanded, and mightily. That is a huge task even on the best of days.

If the size of the dream is unsettling, it seems a colossal mismatch when compared with the size of the company, because all of Sunpower, Inc., fits into an old industrial building in a run-down neighborhood of a tiny Appalachian town. Inside, there is the spareness that comes from twenty years of operating on a shoestring: surplus furniture, scarred floors. There is a human scale to the place that belies the size of the Sunpower dream; a visitor can walk briskly through all of Sunpower in a matter of minutes, or at least through the three gross divisions, those of office, lab, and shop. The offices take up the front third of the building, the lab fills the middle, and the machine shop the back.

The lab and the machine shop are both bordered by many small rooms called test cells, where Stirling machines are assembled and operated. In the machine shop is the test cell for a Stirling cooler, and down a dark hall, past the men's room and the "morgue" (where the carcasses of early incarnations of Stirling engines reside), is the test cell for the biggest machine in the place, the five-kilowatt solar, the machine that might save the earth. In the lab is a test cell for the heat pump project and another for a Stirling cooler aimed at the domestic refrigeration market. In the third is a bearded man in old clothes, tinkering with what looks like some sort of propane gas–powered electric motor. This is the test cell for the small engine, and the man is William T. Beale.

Nick and Chris are still designing the machine, but it looks too good to be true. —*William T. Beale, commenting on the progress of the small-engine group, August 27, 1990.*

Things can *look* too good to be true, but nothing *is.*

This is a consequence of what some people might call "reality," but which engineers know is a consequence of the laws of thermodynamics, laws that *define* reality. The second law, in layperson's terms, specifies that nothing is free, that there is a cost to doing business, no matter what the business is, and that one can never get back as much as one puts in. Knowing that nothing is too good to be true, we are right to be suspicious of, say, a double-your-money-in-seven-days chain letter, just as engineers are skeptical of the engineering equivalent to a chain letter, which is the perpetual-motion machine. The former won't make everyone rich, and the latter won't make more energy than is put in to begin with. Nothing is really free.

There is a particular awareness of this at Sunpower, a place full of engineers who feel the weight of thermodynamic law in every moment of their professional lives. Sunpower, Inc., is a place where everyone is exquisitely aware of the laws of thermodynamics all the time. But it is also a place where this has never prevented hope from springing eternal. William Beale knows you can't *beat* the laws, but he also knows you *can* work at the margin; he is a Philadelphia lawyer of a mechanical engineer, one who knows that it is at the edges of the law that the law is most interesting. And it is at the edges of the laws of thermodynamics that Sunpower works. William Beale has the half-haunted, half-hunted look of a man who has spent the last twenty-five years trying to transform the way the world uses energy.

So William understands that the design of the small engine *can't* be too good to be true, and anyone who knows him as well as the people who work for him do knows that William is an optimist's optimist. They know that what he means is that the small engine appears to be better than he had hoped for and that the small engine might be good enough. Not good enough to beat the laws of thermodynamics, but perhaps good enough to break the bonds of a tiny R & D company and enter what is universally referred to around Sunpower as "the real world." After a quarter of a century of trying to make Stirling engines achieve that leap, he should be forgiven for speaking of something as being too good to be true, and admired for his hope, which is limitless.

To understand a Stirling engine, you will do well to forget most of what you know, or think you know, about engines. Forget oil and gas and (but for a footnote or two) automobiles. If you know anything about Robert Stirling, the nineteenth-century Scottish clergyman who invented the Stirling engine, it is all right to remember this, but keep it in the back of your mind. Instead, think of the Stirling for a moment as a magic box, a box that solves all of the world's energy problems generally, and specifically whatever particular energy problem seems most pressing in a given situation. For example, if the problem is acid rain, imagine the Stirling engine as a magic box that produces energy without producing sulfur dioxides, which cause the acid in acid rain. If the problem is depletion of the earth's protective ozone layer, it is a magic box that makes things cold without the use of chlorofluorocarbons, which deplete the ozone layer. If the problem is dependence on foreign oil for use as a motor fuel, it is a magic box that makes automobiles go on an inexhaustible, cheap, and pollution-free substitute. And so on. Use your imagination. William uses his.

To understand Sunpower, Inc., you should forget most of what you know, or think you know, about corporations. Forget profit and loss and (but for a footnote or two) stockholders. Forget anything you know about large corporations like IBM or General Electric. It does not apply. Instead, imagine a corporation that can fit comfortably inside a conference room, a corporation that meets, in toto, to discuss things like the company cold-drink machine or bulletin boards, a company whose senior financial officer passes out chocolates twice a month to all employees who have filled out their time cards correctly, or even just almost correctly. Imagine a company whose employees meet every Monday morning to talk about how things went the previous week.

To understand William T. Beale, suggests Neill Lane, a senior design engineer at Sunpower, read Paul Theroux's *The Mosquito Coast*. "William is a tinkerer who is out to change the world," says Neill. "He's not dedicated: he's obsessed. It makes him impossible to work for, but impossible not to admire; at the end of the day, William is going to *make* something happen, if he doesn't get killed by someone first."

So three anomalies, at least by real-world standards: Stirling engines, Sunpower, Inc., and William Beale. The first is an invention; the second, a company formed to develop and exploit a refinement of that invention; the third, the man who invented the refinement and started a company to develop the refinement. Add them together on paper, and you have the makings of a truly colossal dream: energy efficiency, for one thing; energy *self-sufficiency*, for another; responsible use of natural resources; utter freedom from pollution—cheaper energy with no environmental costs. A world no longer dominated by petropolitics, a cleaner, cooler, fairer, calmer world. This is how it looks on paper; even on paper, it looks like a tall order for a tiny company in Athens, Ohio.

Thus sayeth William T. Beale. "It would permit us to do the only rational thing, and that is to leave the coal and the oil in the ground for future generations, who will no doubt be far smarter than we are, and who will therefore come up with much better uses for such substances than burning them up and spewing the exhaust into a fragile and declining ecosystem."

The "it" is, simply, the widespread, if not virtually complete, implementation of Stirling technology at the earliest possible moment. When William began Sunpower, the words "greenhouse effect" had only the narrowest meaning and application. "Global warming" was in the lexicon of hardly anyone but a few scientists interested in the phenomenon, rather than in its effects. William Beale's "rational idea" for a Stirling-powered world even predates the 1973 Arab oil embargo, which was the first of a series of wake-up calls America has had regarding energy and the environment. He is not a seer, though, or not only a seer. William Beale doesn't want to see the world turn to clean, cheap, and responsible sources of energy simply because of something so trendy as events in a place like the Persian Gulf. He wants the world to make and use clean energy because it is, to his mind, frugal and sensible and, therefore, simply *right*.

William Beale. Never Bill, rarely Beale, generally William, often William T., or WTB, or William *the*, as in "William the Beale." No one at Sunpower knows exactly what a Beale would be, but it is a name that somehow *fits:* William as a "the," a one of a kind. William—William the Beale—does not look, and seldom acts, like any sort of textbook zealot, which is what it seems someone would have to be to consider changing the world. He looks, in fact, like a gentle, and rather rumpled, college professor, a man at home in his oldest and most comfortable clothes, a man at home at a chalkboard. Sunpower has a half dozen strategically located chalkboards, where ideas and problems are often sketched out, and occasionally solved, and William Beale looks comfortable and kindly in front of them. He looks *natural* there. His brow knits when he concentrates on a problem, and he scrawls with the chalk. There is fire, however, in his voice. On certain subjects—waste of any kind, profligate uses of any substance or resource, activities that are not to his mind sensible or that treat his fellow man and woman with anything less than dignity or respect or, indeed, as anything less than full equals—he goes off like a bomb. (One immediately sees why the federal government comes in for so much of his ire, as does any bureaucracy, any place with more pencil pushers than people doing work.)

Textbook zealots generally have very concrete goals and ideals. They are committed to specific acts. William's passions are more abstract; he is wedded not to the Stirling engine per se but to any solution that would make the world a cleaner and safer place. Stirling technology happens to fill the bill, but he is not parochial about it, exactly. Rather, his passions are channeled toward stopping waste, stopping ruin, stopping exploitation—all intimately a part of the way the contemporary world conceives of and uses energy.

It is not surprising if William's passions often seem to conflict, even if this has the effect of making him seem contradictory, as when he blows up and chastises someone for waste or profligacy, an act that might be seen as treating someone with less than respect. It extends to his distaste for giving orders simply because he is the boss, even about something he sees as unsensible, because this twists him up on the equality issue. He wants things his way, but doesn't want anyone to do anything *because* it's his way. Every day, he begins by looking for consensus, but he also has very strong convictions, which sometimes conflict with the give-and-take needed for consensus.

William Beale's convictions, too, make him seem often to be in a lather, because examples of waste, inequality, and the like are to be found everywhere one looks. Even to watch him read a newspaper is instructive. On some days, it seems as though the *New York Times* and the *Wall Street Journal* and the *Athens Messenger,* a tiny daily full of wire-service stories and pungently conservative editorials, had no other purpose than to set William off; even after all these years, he still can often simply not believe something he has read there. He leans back in his office chair, pulls off his glasses, and rubs his face with his hands, as though he could wipe from his eyes whatever he has just read and cannot tolerate. There are days when he is loath even to pick up the *Wall Street Journal,* so issues pile up for a week or so on the scarred wooden floor of his office, between the university-surplus drafting table and the (famously uncomfortable) "couch," which is really just a wooden bench built into the wall, like a prison bunk. When he does read the paper, as often as not he will see something that begs for comment, demands a response. From the *Journal,* he will neatly clip out the offending article and pin it up on one of Sunpower's bulletin boards, often with a calculation of some kind appended that illustrates waste, profligacy, or shortsightedness, on the one hand, or penurious-ness, blindness, or inattention to the knitting, on the other. "The total amount of money spent on energy research in the United States equals one third of one cent of the price per gallon of gasoline" is one example, stabbed with a blue pushpin to a *Journal* clip on talk of a fifty-cent-per-gallon gasoline tax to pay off the federal deficit. There is always the tone of amazement and injury, the tone of a man who has suffered the slings and arrows of outrageous fortune until he is humpbacked, and still the outrages continue to accumulate. To the *Messenger,* he can often not resist the impulse to write. "Like many readers of this newspaper, I have many things I would rather do than write (or read) letters about the tragedy in the Persian Gulf," begins one crisp missive to the *Messenger,* a letter that dates from the 1991 war, but that might also begin a letter written in 1981, or 1971, because to William the continuing tragedy of the Persian Gulf is not only war but also what he sees as underlying all war: war, he likes to say, is a primitive failure of imagination. His own imagination is fertile, a continuous wellspring of thought and idea and sensibility, concerned generally with the practical and particularly in the Stirling engine. This engine is not the answer to everything, but it is the answer to many things, and hence would permit the nation to devote its energy to more recalcitrant problems than how to fuel itself. The answer to that question is *simple*.

Someone who was less single-minded, less resolute—that is, virtually everyone else—gives up or gives in to a certain degree and stops trying to change the world. William Beale has not done so; instead, he started and has sustained Sunpower, Inc., a company now more than twenty years old and an odd monument to his ideals. Because if there is a Supreme Being, it is a Supreme Being with either a sense of humor or a sense of pathos, or probably both, a Supreme Being that admires the labors of Sisyphus, a Supreme Being that gave to William Beale something that looks as though it could and should fundamentally change the world and then, rather than provide the resources necessary to implement the change, spreads those resources among the defense industry and Wall Street raiders, among government agencies that react so slowly to change that simply maintaining the status quo seems like a daring move.

The Supreme Being giveth, and the Supreme Being taketh away, and what the Supreme Being giveth in the case of Stirling engines is a mechanical device that to a layperson sounds as though it could have been invented only by a Supreme Being: a Stirling engine is, theoretically at least, not only more efficient than the engines and power sources we have come to depend on but a power source independent of fossil fuels, like petroleum or coal, and a power source that, indeed, works best using the "free" energy source of the sun. Oh, yes, and the Stirling engine also makes what is, for late-twentieth-century purposes, a perfect refrigerator and air conditioner: more efficient than the conventional devices and operating without the chemicals that are destroying the earth's ozone layer. The same device solves two of the fundamental problems of the industrial age; by providing all the power we can use, without harming the environment, and all the cool we want, ditto. Good-bye greenhouse effect, good-bye holes in the ozone layer, good-bye geopolitical chess games in the Persian Gulf (as William likes to point out, the Saudi Arabians have more sun than oil). Good-bye global warming. If William were in charge of the world for a few months, he would make it a fundamentally different place. This is the idea, the hope, that begat Sunpower, Inc., a tiny company in a tiny town in an out-of-the-way corner of Appalachian Ohio. Stand here with a long enough lever, and the world will *move*.

William Beale has the idea—the place to stand—but he has heretofore been unable to get a good grip on the lever. Sometimes it seems like fate. At other times it seems a conspiracy. For every handful of R & D money that Sunpower manages to attract, shovelfuls are thrown at everything else, and truckfuls are simply wasted. During the Gulf War, the currency of comparison around Sunpower became the weaponry of destruction. "For the cost of *one* Patriot missile . . . ," "For the price of one-tenth of a Stealth bomber . . . ," "For the amount of money it costs to run an aircraft carrier for one hour. . . ." At a Monday morning meeting, Al Schubert suggested, tongue-in-cheek, that one of the Stirling refrigerators in development could be sold to the military as a machine-gun cooler. The laughter was bitter, and ironic. Following the spill of crude oil by the *Exxon Valdez,* the currency calculations were based on the cost of the cleanup; when several oil companies announced stratospheric profits the following quarter, the currency conversions were based on smaller and smaller fractions of those profits: "With the profits from *one day's* use of gasoline. . . ." Sunpower is a place where technology, where research, aspires to be pure, but also a place that needs to be tainted, just slightly, by the odors of capitalism. Instead of taint, it generally gets taunt. "Q: What makes a Stirling engine run?" one reads on a Sunpower chalkboard. "A: MONEY." Sunpower's struggles for money touch every one of its activities, every employee, every purchase and negotiation and contract and expense.

It shouldn't be surprising that there are some start-up costs associated with changing the way the world uses energy, but this principle seems lost more often than not on the people whose interests Sunpower is working to protect—that is, everyone. Instead, the response of the world at large to Sunpower is typified by a response Sunpower received from a huge international manufacturer of refrigerators. Refrigerator manufacturers have been told by the world to find an alternative to chlorofluorocarbons, or CFCs, which are damaging the earth's protective ozone layer; a Stirling-cycle refrigerator uses *no* CFCs. As sort of a bonus, it turns out that Stirling-cycle refrigerator compressors are *more* energy efficient than the conventional vapor compression technology on which refrigerators and air conditioners and heat pumps are based now.

Sunpower courted, long and earnestly, this international refrigerator manufacturer, a company based in South Korea. There were exchanges of letters and drawings and specifications and proposed contracts; there were many exchanges of faxes. There was a visit to Sunpower by engineers from South Korea. More faxes. More negotiations. And finally, a response, announced at Sunpower by John Crawford, the senior engineering manager whose job it is to drum up prospective sponsors for development of Stirling technology. The South Korean manufacturer had "decided that they could exploit conventional vapor compression technology for another decade or two," John said at a Monday morning meeting, "so, that's that." Although the word "exploit" was most probably used innocently, it is the sort of picture that Sunpower was used to seeing: the world may be ready for Stirling technology, but no one wants to pay to make Stirling technology ready for the world. Particularly if the world can be exploited for another decade. Or two.

Once, in the early 1980s, William took a Stirling water pump to the offices of the Agency for International Development, a body ostensibly concerned with international development. The problem of pumping water is not trivial in the Third World; without cheap, simple, and reliable water pumps, water is carried, by hand (and, in many countries, by *head*; witness those pictures of village women with clay pots balanced atop themselves) from pond or spring to village. William has been trying to get water pumps to the Third World for two decades. The pump he brought to AID was something that he saw as destined to make better the lives of millions in the Third World, by providing a cheap and simple method of moving water. He fired it up in the office of the director. "They told me," William says, in the tone of voice familiar to those who know him as the tone he uses when he is about to speak of an atrocious frustration, "that I was getting water on their carpet. It was a *water pump!* I told them to quit worrying about the carpet and worry about the water, but they were more interested in the floor than they were in the technology, and the sooner I was out of there, the better. That's the way that bureaucracies think. They care more about the carpet than they do about the people they pretend they are trying to 'help.'"

That, in a nutshell, is Sunpower and the world: the world—the real world—is one sort of place. Sunpower is another sort of place, where no one worries about the carpet, carpet symbolizing ornament covering function. Sunpower was conceived of as a functional organization in the extreme, where fine-tuning of an important idea would occur and where the result of that tuning would be, like all fundamental technological advances, invisible except in its ubiquity. But an idea is only half of progress: without implementation, an idea has no worth except as intellectual stimulation. Sunpower is where an idea is supposed to be implemented, so that progress may occur. To date, the company is exactly halfway to progress.

At some point, a question arises, much like the sardonic question asked of economists: "If you're so smart, why aren't you *rich?*" If the Stirling engine is so *good,* why don't we have it already? The two short answers, closely related, are appropriately macroeconomic. One answer is "supply." The other is "demand."

The bald fact is that the Stirling engine, except for a time in the nineteenth century, and for several highly specialized uses, has never really caught on: there is no infrastructure of Stirling technology, no critical mass of researchers or research dollars devoted to development. The Stirling has only tangentially ever appealed to the military-industrial complex, which is where a huge fraction of American research dollars go, and because it is both an "alternative" technology, which means it is something we don't have, or make a lot of, already, and because it is powered by "alternative" fuels, for which no infrastructure is in place, it has no built-in constituency. All the other power sources we use, from automobile engines to nuclear power plants to hydroelectric dams, have a sort of fraternal organization, complete with customers and suppliers and lobbyists and strategists. Stirling engines have none of these. Jan Meijer, an engineer who headed the Dutch corporation N. V. Phillips's Stirling-engine program for years, once answered the question "What's wrong with the Stirling engine?" by saying, "The existence of other engines."

This is not exactly a conspiracy theory, such as the one that suggests that the technology for a one-hundred-mile-per-gallon automobile is available but not accessible, because of an unholy alliance between the oil companies and the automobile manufacturers. Would that it were something so dramatic! Stirling technology is stifled by more-prosaic factors: there is no, or little, demand, because there is no, or little, supply; there is no supply, because what a Stirling engine produces—usable energy—is already produced by a highly developed and mature set of industries, all well established and all interdependent. Any incentive for change in such a highly evolved field as energy production would have to be extremely powerful. William Beale and the people who work at Sunpower believe that the incentive is both real and powerful—nothing less than the survival of the planet. The situation would be quite different if, say, the presidents of the five hundred largest corporations in America believed that the incentive is real and powerful.

For Stirling technology to reach some sort of critical mass, something else must happen. The company got a boost of sorts in the 1970s,
when there was a push for exploration of efficient and clean technologies. But when the price of oil came back down, the interest in alternatives flagged. Other windows of opportunity have been opened, ever
so slightly, which makes them easy to close when times, priorities, or
people change. The Stirling-technology industry has never grown to
anywhere near the size it would need to be in order to either attract the
big players or be, like some military suppliers, too big and too tied to
the national interest to be permitted to fail. That the technology is close
to coming to fruition is not necessarily an incentive to the energy
industry; rather, it is only a frustration to those who believe in the
technology. People like to say that the free market determines the *good-
ness* of new technology; if something is good enough to survive, it will
survive and prosper. And if the playing field were level, this might be
true. But the U.S. government has never seen fit to call out the military
to protect the raw fuel for Stirling engines, as it has for other fuels,
nor does the government build, at taxpayer expense, the infrastructure
needed for the technology to thrive, as it does when it builds roads and
highways and bridges for the automobile industry. It does this because
everyone who owns an automobile is a beneficiary of these endeavors,
and thus collectively a powerful motivator of endeavor. This obvious
fact spawns the part rueful, part resentful currency calculations of how
many Stirling engines could be developed for the cost of one weapon
of destruction or gas-guzzling widget. Stirling technology has, at this
time, no such constituency: if all the believers, the true believers, in the
technology were magically to relocate to one place, it's doubtful they
could elect a representative to Congress. The Stirling engine is thus
part orphan, part neglected stepchild—the worst combination when it
comes to effecting fundamental change.

Unless, of course, the needs of the culture change and what a
Stirling engine does becomes desirable and important. In a perfect
world, this would happen for Sunpower: the world would suddenly
beat a path to its door. William Beale is at least willing to meet the world
halfway as it discovers what carbon dioxide does to the atmosphere.
Meeting the world halfway, though, should not be seen exactly as a
benevolent act. He just happened to be aware of a growing problem a
couple of decades before everyone else.

The whole idea of Sunpower, Inc., is predicated on a firm belief
that the Stirling engine offers solutions to problems that contemporary
society finds more and more vexing: pollution of the air and water and

land; profligate use of natural resources, particularly fossil fuels like oil and coal; the waste of energy through the use of technologies that were perfected when the air was clear and the costs, both real and projected, were low. That all of these actions go against William Beale's personal grain is simply coincidence; that the world is suddenly becoming aware of what millions of tons of carbon dioxide are doing to the atmosphere we live in, and what millions of pounds of CFCs are doing to the ozone layer that protects us, is possibly fortuitous. A generation ago, this country was only beginning to pay attention to the fragility of the planet; now it looks as though the world might be ready to listen— even just a little—to William Beale.

A simple example of this would have been too farfetched to posit even a generation ago. Imagine a Stirling engine (forget for a moment that you don't even know what a Stirling engine is) that uses the same amount of energy as a household furnace in winter, and the same amount of energy as a central air-conditioning system in summer, but that generates two things: one is enough electricity to meet the needs of a typical household—to disconnect the house, except during times of peak use, from the grid of wires, transmission stations, power plants, coal mines, nuclear fission reactions, dams that block scenic rivers and the like; the second product of the Stirling in the basement is heat, which a Stirling engine needs to reject anyway. A Stirling in the basement that generated enough electricity for the household to function would reject—give off—enough heat to keep home, hearth, and hot-water heater humming along comfortably in winter, enough cool to do the same in summer. It sounds implausible, yes? Retrofit every American house with a box in the basement that generates free electricity?

Only fifteen years ago, though, "heat pump" was a term meaningless to all except engineers who knew their thermodynamics. Nowadays, a heat pump is what is installed in about one out of three situations when home owners think they're buying a central air-conditioning system. Heat pumps evolved from being equations in mechanical-engineering textbooks to consumer items because electric companies saw a way to make a buck. It is, in most places in the country, an invisible technological advance—hot air comes out in the winter, cold air in the summer, and who cares where it comes from, or why. Imagine, now, a heat pump that has no CFCs, runs on natural gas in winter *and* summer, produces less pollution than typical furnaces do, runs more efficiently. Imagine that it also produces most of the electricity you use. This sort of picture is what draws some engineers to the Stirling engine.

It is still only a picture, however; there exists no magic box at

Sunpower that does all of these things. What Sunpower has is the idea, and the rudiments of the technology, to make such a magic box. As important to note as their technological progress, though, is the scale of their dreams, and the implications of those dreams. Like William Beale, when talking about the promise of the small engine, they envision "a box in every basement" with a Stirling engine inside, making electricity from waste heat. *Every* basement—just as Neill Lane dreams of a desert full of thousands of solar dishes, waking up each morning with the sun, shrugging their shoulders, and making electricity, just as Lyn Bowman, whose Stirling microcooler is little more than a hundred pages of thermodynamic equations at the moment, envisions a time when microcoolers are ubiquitous. The engineers at Sunpower are not the sort who look to squeeze an extra twentieth of one percent efficiency out of a household refrigerator, or develop a device to knock twenty-five cents off the monthly electric bill. The engineers at Sunpower are out to remake the world as we know it, and to do so by means of the Stirling engine.

They spend most of their time on the technological issues, because the technological issues are the most immediately and concretely pressing and must be addressed fully if Stirling technology is to have any chance at all. But anyone who suggests that William Beale and the band of merry souls at Sunpower don't have a political agenda is seriously misreading that look in William's eyes. "As we stand at the beginning of the last decade of the 20th century," William Beale wrote, not too long ago, "we suddenly find a bright new world opening before us. The specter of nuclear annihilation, which has haunted us for half of this century, is now faded in the dawn of a new global possibility for peace and cooperation. And just in time! With the threat of instant death finally lifted, we come face-to-face with a slower, but equally fatal threat—environmental collapse. Under the pressures of exploding human population, ever-rising demand for material goods and energy consumption, devastating third-world poverty and consequent rapine of the still-remaining forests and soils, the thin, fragile ecosphere upon which all life depends is under the gravest threat."

This is pretty snappy copy in almost any context, but as a hook for selling Stirling engines, it is without peer. There is a Stirling engine lurking in there (the quotation is from a paper entitled "The Promise of Solar Thermal Energy Conversion"), not for the beauty of the technology, or as a move in a geostrategic game played out in Kuwait or Iraq, but as a way of avoiding a path to certain ruin, a way, it is not overstating it to say, of saving the world. In geopolitics, this might be seen as a transparent attempt at staking out the high ground. At Sunpower, this is the only ground that William Beale will even consider staking out. In concert with the technology it is developing to retrofit the world with Stirling engines, such plans make Sunpower a very interesting place.

Chapter 2

The Contraption in the Stairwell

Natural processes are accompanied by an increase in the entropy of the universe.

—the second law of thermodynamics

You think you're in trouble now? Just you wait.

—Neill Lane's version of the second law

NEILL LANE reaches with one hand for a piece of chalk and with the other for a cigarette, and as he begins to speak of thermodynamics also begins to draw the fundamentals of a magic box. All three actions—drawing, smoking, and speaking of thermodynamics—are habitual, almost instinctive, and all are as much a part of Neill Lane as, say, his sense of humor or his extravagantly curly hair. His impulse when he is faced with any question is to go to one of Sunpower's ubiquitous chalkboards, and sketch it out, his favorite chalkboard being the big green one in the machine shop, because he can then have a cigarette at the same time. The machine shop is the designated smoking area at Sunpower.

When he is anywhere but at the chalkboard in the machine shop, he can't reach for a cigarette, since smoking, like wasting energy, falls into the category of frowned-upon activities at Sunpower. Indeed, the company "ashtray," which is actually a steel drum filled with sand just inside the swinging doors that lead from the machine shop into the rest of the building, is situated in such a way as to permit all of the nonsmokers to pass by regularly, often with a wave of the hand to disperse the smoke. You can have a cigarette at Sunpower, but you generally have to have it in a place that enables all the nonsmokers to observe, disdain-fully, your weakness. It's like coming to work in the winter without your sweater, or using the laser printer for anything other than im-portant business.

So if Neill is in the drafting room, or his office cubicle, he may *hold* a cigarette, but he never forgets and lights it. Instead, he will rustle around on his (famously) disorderly desk for his most recent coffee cup, and sip coffee while he sketches. What he draws when he's asked how a Stirling engine works looks not like an engine but, rather, like the *thermodynamics* of an engine. When pressed (or when giving his cocktail party description of what he does for a living), he will draw a piston in a cylinder, but that's about as far as it goes. It's the *thermody-namics* that make engines go, and it's the thermodynamics that draw engineers like Neill Lane to the Stirling cycle.

Neill himself was drawn to Athens from the University of the Witwatersrand, in Johannesburg, South Africa, where he was a junior lecturer in thermodynamics, a field in which by all accounts he has a special competence; Sunpower is a place where thermodynamics are real, where a thermodynamicist's dreams can come true. (Another engi-neer at Sunpower calls them "technical fantasies.") That it is also a place where such dreams can be shipwrecked goes without saying; to make a technical fantasy come true, one has to do more than simply *have* it. One has to make it into hardware. Neill is notoriously dismissive of any idea that is only imaginative, of any idea that can't be made real; he is impatient with meandering sentences that begin, "You know, you *might* be able to . . . ," which is one Sunpower mantra, or "If we could . . . ," which is another. Neill always interrupts and says, *"How?"* He lives in the real world, thermodynamically and otherwise. "Require-ments, constraints, criteria" is what Neill used to tell mechanical-engi-neering students at Wits, and what he is always saying at Sunpower. "Requirements: what is the machine supposed to do? Constraints: what are the limits you have to work within? Criteria: what are the parameters of the design?"

In other words, if an engine is supposed to put out five kilowatts, it shouldn't be designed to put out six or four; if an engine has to fit into a suitcase, it shouldn't be conceived to be the size of a steamer trunk; if an engine is supposed to run off natural gas, it shouldn't be configured with a solar dish. You give the customer what the customer wants, not what you think he needs. In this, Neill is sometimes the complete antithesis of William, who often finds sponsors to be timid, shortsighted, and unadventurous, and who thus feels it's his *responsibil-*

ity to give the customer what he thinks the customer should have. This is the ultimate in technical fantasy, to give the world what you think it needs. In *Sunpower Newsletter,* number 1, which dates from January of 1974, William Beale wrote, "What *I* want Sunpower to be getting toward are things that people need even if they don't know they need them yet."

But Neill doesn't want to make the world in his image: he wants to see a factory producing Stirling engines, he wants to see an assembly line cranking them out, he wants to see them lined up in the desert at sunrise, waking up and making electricity. He wants a production machine. And he believes you get this if you know, intimately, the requirements, the constraints, the criteria. If Neill Lane has a slogan, it consists of these three words. So when he steps up to the chalkboard and sketches a thermodynamic cartoon of a Stirling engine, he is sketching what is only the starting point, the theoretical ideal, the place where all Stirling engines are the same. After this, they are all different, depending on the requirements, the constraints, the criteria.

He is able to begin with this sketch, this common denominator, because the Stirling cycle is functional, as all basic technologies are functional, and at bottom is as plain as a brick. The Stirling cycle is as simple to someone who loves thermodynamics as it is opaque to someone who doesn't. Technology is interesting in this way; at its purest, it is objective. A device, an invention, a machine, works according to the laws and principles of nature, according to the laws of thermodynamics; devices that don't work according to these laws and principles don't work, period. How well a machine works may be measured in absolute or relative terms, but it can be measured. There are good designs and bad designs, but designs are good and bad depending on the requirements, the constraints, the criteria. One broad example of this is the 5K machine itself, which has for all practical purposes *no* weight constraint whatsoever: it is intended to sit out on the focal point of a solar concentrator, a curved dish that will focus the sun's rays, and needs to have a mass heavy enough to balance the mass of the dish itself. It's like trying to counterbalance the offensive line of a football team on a seesaw. So the 5K is bigger, heavier, than it might be, because from Neill's perspective, there is no advantage whatsoever in trying to minimize weight: no matter how heavy the prototype ends up being, it will still need an additional counterweight of some kind, so why shave a few pounds off the weight of the engine?

On the other hand, had the engine had a strict weight constraint, it would have been envisioned and designed very differently from the start. Neill bristles like a porcupine when anyone at Sunpower comments on the 5K's size, because he almost takes it as an affront, as though he were too stupid to have noticed that the engine was a large and heavy object. "There's *no weight requirement*," he says, over and over, when Robi or David pops in and points out that the 5K sure is *big*. The 5K, big or no, is simply an envelope for the thermodynamics, and Neill Lane seems sometimes to be cursed by people who ought to know this better than anyone. It is the thermodynamics that make an engine *go*, but the requirements, constraints, and criteria make an engine, and make all engine designs different one from another.

So what Neill Lane draws first—it's called an "indicator diagram"—is exactly what any engineer might draw when puzzling out how an engine might work, or at least any engineer since the middle of the seventeenth century. It is exactly the sort of drawing that James Watt, the inventor of the steam engine, kept as a trade secret, because it illustrated to those who could understand it just how, and why, a steam engine would work.

To understand an indicator diagram, though, requires some background. Engineers know this background; everyone else uses the implications of this background every day, often without understanding it in the least. This is when technology—the automobile, the electrical outlet, the air conditioner—is absolutely and completely opaque: you can look all day at an electrical outlet and never see the power plant behind it; a car goes because you put gas in the tank and turn the key.

Such an invisible infrastructure is one thing that contemporary technology has wrought exceedingly well. An infrastructure like ours encourages an *opacity* of technology: we fill our worlds with devices, with machines, that are aimed at being "user-friendly," which more and more means "user-ignorant." Press a button marked "on," and the VCR, the CD, the PC, comes on. Turn the key, and the car starts; pull the cord on the weed trimmer, and it whines into action. Somewhere, in some research lab, someone (an engineer, in fact) thought all of this through and presented the consumer with a device altogether opaque. "No user serviceable parts inside" is a sentence describing one of the fundamental achievements of late-twentieth-century research and development, so that we don't dare take the back off the CD player, for fear that the magic will escape.

This is the direction of machinery as mystery. One art of the engineer is to figure something out; the other art of the engineer, like the art of the poet, lies in concealing the art. Both are as much a part of the world today as, say, cheap gasoline and cheap electricity. You press a button, and the magic comes out.

This is not necessarily so, though, at the level of invention, of development, of prototype. Research and development is the stage in the development of technology when the back cover is off and all the inventor-serviceable parts are exposed to the scrutiny of the inventor. Stirling engines have come a long way in recent years, particularly at Sunpower, but they have not come along so far as to be technologically opaque. Not many people know what makes a CD player go (not many people, for that matter, know where the electricity comes from that makes the CD player go, either), but they know that someone has figured it all out in such a way that it works. Sunpower is the sort of place where the figuring out goes on, where inventions are conceived, developed, refined. Someday, perhaps, Stirling engines will be as technologically opaque as the internal-combustion engine (like that under the hood of an automobile) or the nuclear power plant (like that next to those Three Mile Island–style cooling towers). Then a consumer will not be asked to marvel at the notion that the electricity that runs the air conditioner, or the air conditioner itself, is based on technology that flowed from the imagination and optimism of William Beale. Between now and then, the technology must be made.

So the first thing to recognize about the Stirling engine is that it is, at least, very unfamiliar, which makes it seem a bit complicated to laypersons and, for that matter, many engineers, most of whom studied the Stirling cycle for an hour or so, if at all, as undergraduates. For every Stirling engine in existence, there are literally millions of, say, internal-combustion engines, and many thousands of conventional power-generating plants, and engineering curricula, like most curricula, stress what the world has, rather than what it lacks.

But the Stirling cycle *is* fairly simple, at bottom. A Stirling engine belongs to the class of engines called heat engines, which includes any contrivance for converting heat into mechanical energy; the engine in an automobile is a heat engine, too. However, the Stirling is an "external-combustion" engine, in that the heat source that powers the engine is outside its working space, a working space made up of a piston and a cylinder. A steam engine is an external-combustion engine, but an automobile engine is not: in an automobile engine, the burning of fuel that makes the car go occurs *inside* the cylinders; hence, it is an "internal-combustion," or "IC," engine. It is because combustion takes place inside the cylinders of an IC engine that its fuel requirements are so relatively strict. In a Stirling, the combustion is outside, so the heat can come from the combustion of anything—wood chips or cow chips, for example, or the focused rays of the sun. It is important to note a quality of external-combustion engines, a quality shared by the Stirling and the steam engine: combustion is continuous, rather than episodic, as in an IC engine. Steam locomotives had a firebox in which a fire burned continuously, heating water inside a boiler to steam.

So when the heat source is outside, rather than inside, the cylinder, the choice of fuels, and the quality of the fuel, becomes far less restrictive than in an IC engine. The choice becomes, in fact, nearly limitless. Furthermore, continuous combustion, rather than the series of timed explosions that occurs in an IC engine's cylinders, changes the character of the engine profoundly. Both processes convert heat into mechanical energy, one by continuous application of heat, the other by episodic combustion. But even when the amount of heat generated is equal, the form it takes is quite different—one can heat a room by setting off a succession of firecrackers inside a wood stove, or one can throw a log or two inside, and sit and read the newspaper while they burn.

Continuous combustion also permits far more *complete* combustion, which has the practical effect of improving the efficiency of an engine, by getting more energy out of the fuel. A combustion process that burns only 50 percent of its fuel can be thought of as "wasting" half of the potential energy of the process, waste that takes familiar form as, say, automobile exhaust. (An inefficient engine also wastes energy in other ways, of course.) No combustion process is perfect; all are, to a greater or lesser degree, inefficient, and what we call "pollution" is the result: pollutants are better thought of as unburned, unused, fuel. So an external-combustion engine that permits the use of extremely efficient combustion processes is "cleaner." But from Sunpower's point of view, it is more important that external combustion permits the use of heat sources that do not involve combustion processes at all, such as the focused rays of the sun, and that thus are as clean as clean can be.

So a Stirling engine turns external heat from a continuous heat source into mechanical energy, whereas the most familiar heat engine, the IC engine, turns internal heat from episodic combustion into mechanical energy. To schematize the two engines this way implies that both have certain advantages and disadvantages, but both, of course, work.

But how *differently* they work! To a trained eye, the contemporary Stirling engine, indeed, seems to be the very opposite of the IC engine.

One uses what is, finally, an exotic and rarefied fuel, a fuel that is geopolitically volatile and environmentally harsh; the other uses any fuel, the more geopolitically benign, the better. One produces exhaust emissions and pollutants that foul the air and damage the atmosphere; the other produces much smaller quantities or, in certain circumstances, none at all. It would fit the pattern to say that one is a hoary and antiquated nineteenth-century design, while the other is a product of the new age; but this statement would not be true. The Stirling, in fact, predates the IC engine by nearly half a century.

Robert Stirling took out his first patent in 1816. At that time, there were machines that harnessed an existing mechanical force, like waterwheels or windmills, and there were steam engines, which converted heat into work by means of a mechanical cycle. And that was it.

The original beauty of the Stirling engine, circa 1816, was that it held out a way of providing mechanical power at low pressures: the early 1800s marked a sort of cusp between the Dark Ages of mechanical engineering and its Renaissance, a period when a better understanding of thermodynamics led to inventions of all sorts, but inventions that often outstripped the science of calculating the strength of materials. Early steam engines were terrifying machines because of this; the hotter the fire, the higher the steam pressure, and the faster the engine will go. Until the whole thing explodes, that is; even steam engines operated conservatively were subject to failure, since an understanding of metal fatigue was at that time in its infancy; anyone who has broken a wire coat hanger by bending it back and forth a few times has an inkling of metal fatigue. It also turns out that temperature cycles—hot, to cold, to hot, to cold—have an immense and important effect on material fatigue; while this was often appreciated after the fact in the early days of materials science, it was a long time before anyone could do anything about it, or even envision a way of testing it.

Robert Stirling was not a materials scientist; he was a clergyman. P. W. Atkins, a physical chemist at Oxford University, points out that Stirling, as befit his calling, grieved over the tragedies caused by the fairly common explosions of steam engines. Superheated steam is one of the most horrible of substances, melting the flesh off anyone unfortunate enough to come into contact with it; and it combines with the shrapnel of an exploding cast-iron boiler. As a result, death and maiming by steam engine were all too common costs of the early Industrial Revolution. So Stirling set out to devise an engine that would operate at low pressures, and for a time in the nineteenth century the Stirling engine was a popular and widely used prime mover.

It had two disadvantages, even by nineteenth-century standards. First, the engines were very large relative to the amount of power they produced; second, they were voracious consumers of fuel—in engineering terms, they had a low power density and were extremely inefficient. Still, they offered an appropriate technology for certain times and places, such as coal mines, where there was no shortage of fuel and where size didn't matter. They were also exceedingly safe, in that they operated at or near atmospheric pressure; when they failed, they broke down, but they did so with a whimper, rather than a bang. The comparative weaknesses of the Stirling, and the combination of a better understanding of metallurgical science and the invention of the IC engine, relegated the Stirling to a few specialized uses, the experimentation of hobbyists and enthusiasts, and engineering textbooks, where the Stirling cycle is still taught, if only briefly.

Until the time of William Beale. How the Stirling has come to hold—for the people at Sunpower, anyhow, or at lest for now—the embodiment of the future of the planet is quite a story, which needs some background first.

The Stirling cycle, simplified version, is this: an inert gas (helium, hydrogen, or, as in Stirling's Stirling engines, simply air) in a cylinder is alternately expanded, cooled, compressed, and heated in a continuous cycle; the cooling comes in part from what engineers would call a "cold sink," which is the generic name for any device capable of removing heat from a system—if a bottle of wine forms a system, the ice bucket is a cold sink; in an automobile engine, the cooling system acts as a cold sink. The heating comes in part from any heat source—a burner, like the burner on a gas stove—and the expansion and compression are accomplished by a piston inside the cylinder.

In a Stirling engine, the gas that does the work is not the fuel but, rather, simply a working fluid, and it is recycled continuously—in Sunpower engines, the engine is hermetically sealed inside an airtight canister called a pressure vessel. The Stirling also has a continuous heat source and a continuous cold sink, and uses them to the advantage of the engine, and both are outside the cylinder of the engine, rather than inside. In other words, while a heated gas moves a piston in a cylinder and does what an engineer would call "work" in both external- and internal-combustion engines, in the Stirling the gas is cooled and heated again, in a cycle, rather than evacuated and replaced. So the power stroke of the piston is, for the purposes of understanding how the devices work, similar in the two engines. It is the other strokes that are different.

To understand how, as well as to get a sense of why the difference is so important, one doesn't need an engineering degree or even mechanical aptitude; one needs only a single fact from elementary school science; the rest is application.

The fact is that "a gas expands when heated, and contracts when cooled." Blow up a balloon, and put it in the refrigerator. In twenty minutes or so, take it out. It will have shriveled to a greater or lesser degree (depending on the elasticity of the rubber in the balloon and the temperature in the refrigerator), but it will have shriveled some all the same. And as it warms up, it will expand. Gas expands when heated and contracts when cooled.

There are two corollaries to this, corollaries that explain why Stirling engines go. The first is that elastic containers like balloons permit the gas inside to reform, to reshape, their enclosure; a rigid cylinder made of, say, steel does not. So any gas in a rigid cylinder increases in *pressure* when it expands, and decreases in pressure when it contracts: heat increases pressure; cold decreases pressure. This is pertinent for any sealed container or vessel: the gas inside it expands when heated and, having nowhere to go and nothing to move, increases in pressure.

The second point is subtle, but obvious. An increase in pressure raises temperature, and a decrease in pressure lowers temperature, two sides of the same coin, really. But someone versed in thermodynamics might say, "A gas gives off heat when compressed and absorbs heat when expanded," another way of putting the same fact in thermodynamic terms.

The laws of thermodynamics tell us, simply, that energy is not created or destroyed but, rather, only changes forms: the amount of energy in the universe is fixed, and when a change occurs—like a change in the pressure of the air in a balloon, the energy has to *go* somewhere. When a gas, like the air in the balloon, is expanded, it absorbs heat; when the air in the balloon "cools," the pressure drops.

This is, for example, how an air conditioner or a refrigerator works, by containing some substance under pressure that absorbs heat when it is expanded and gives heat off when compressed. Unfortunately, air-conditioning and refrigeration are perhaps the two most opaque technologies ever perfected by a culture. A compact disk looks something like a record; a computer looks something like the combination of a television and a typewriter; a refrigerator is a magic box. But the "magic" is simple, really; the substance in a refrigerator's coils is some form of chlorofluorocarbon, a substance with a boiling point much lower than the temperature in, say, a house—or in a refrigerator.

The air in the house is always trying to warm up the inside of the refrigerator, and the working fluid inside tubes in the walls of the refrigerator is always absorbing this heat, because the working fluid has a low boiling point and is consequently expanding, and a gas that is expanding absorbs heat. If the working fluid were water, it would have to be above its boiling point for it to be a gas, and to get it above its boiling point, a lot of heat would have to be put into it, just as a lot of heat has to be put into a pot on the stove to get water to boil. (There are lots of other reasons why water wouldn't be a good working fluid in a refrigerator, but this one alone will do.) The heat source in a refrigerator, to simplify, is the empty space inside; an air conditioner is like a refrigerator with its door open. Compressing the working fluid outside the refrigerator makes the working fluid give up its heat into the kitchen (those tubes on the back of a refrigerator contain the compressed working fluid, which is why they are warm; this is also where all that hot air comes from on the outside of an air conditioner).

A subdivision where everyone has an air conditioner can be seen as a "room"—the subdivision—full of refrigerators; interestingly, places like densely populated suburbs and cities with lots of air conditioners actually *raise* the temperature outside by fractions of degrees as the cost of cooling the interiors of houses and buildings. An engineer would point out that the air conditioners are simply "pumping" the heat outside. This is a nice and simple illustration of one of the effects of the first law of thermodynamics: heat, like any other form of matter or energy, can't be destroyed, dispensed with, made to vanish; it can only be moved around. This is what physicists mean when they say that the amount of matter in the universe is fixed, and cannot be created or destroyed.

The second law—the law that has everything to do with why heat engines work—is the logical extension of this: the amount of energy may not change, but its form *does* change. And it always changes in one direction, that of disorder. This is entropy, the changing of energy from a higher order—say, a log in a fireplace—to a lesser order—ashes and smoke, and, of course, heat. The second law explains why a hot object cools, but why cool objects don't spontaneously get hot; it explains why a bouncing ball comes to rest, but why balls at rest don't simply begin bouncing of their own accord. A second-law engine, then, is one that moves heat around in such a way as to get some work out of the process, but in the big picture contributes, however slightly, to the decay of the quality of the energy in the universe. An automobile engine is perhaps the perfect example of a second-law engine, in that it takes higher-order, high-quality energy—oil—and turns it into whiffs of pollution. The first law says that all of the energy in the universe exists in a fixed amount; the second law tells us that no matter what we do to the energy in the universe, one effect will always be to degrade the energy from a higher quality to a lower quality. It is always all downhill from here, no matter where here might be.

One of Aesop's fables hints at an illustration of thermodynamics. A weary traveler on a cold winter's night stops at a lonely house and knocks on the door. The man inside admits the traveler, commiserates with him over the brutal weather, and offers him some soup. The traveler is grateful and walks over to the fire to warm himself, blowing on his hands as he does so. His host asks, "Why do you blow on your hands?" The traveler answers, "Why, to warm them with my breath." A few minutes later, the host serves a bowl of soup ladled from a steaming caldron, and the hungry traveler blows on the soup. "Why do you blow on the soup?" the host asks. "Why, to cool it off so I may eat it," answers the traveler. "OUT!" roars the host. "Out of my house! You must be the devil himself, able to blow hot and cold with the same breath!"

Thermodynamics is like this, in that "hot" and "cold" are relative terms, best used only to describe a temperature differential, rather than a state of being. To "cool" something is really to take heat out of it temporarily and put it somewhere else—into the kitchen, into the backyard, into the atmosphere—because you can't simply *get rid* of heat. You can only move it around. Heat is everywhere inside the atmosphere, and we cool things by moving the heat around. A Stirling engine is a particularly efficient device for moving heat around, whether we are using heat to do work, or using work to pump heat, or, if we pair a Stirling engine and a Stirling cooler, to do both. A house, after all, has hot uses and spaces—a furnace, electricity, the hot-water heater—and cold uses and spaces—the refrigerator, the air conditioner. Put the right combination of parts in the box in the basement, and the thermodynamics of a household come into equilibrium. This is extremely efficient. That is why the idea appeals to William Beale.

But what's in the box? There are Stirling engines, Stirling engines, and Stirling engines. The first is the theoretical cycle, which forms the foundation by which the second—machines based on Stirling's original design—and the third—William Beale's refinement—operate. The first is what Neill draws on the blackboard; the second is what the hardware looked like until the mid-1960s; the third is what Sunpower makes.

Second things first, which may be appropriate, since there is some evidence that Stirling himself didn't understand his own machine, at least not initially. This is not unusual; it may be that Nikolaus August Otto, who "invented" the IC engine, and Rudolf Diesel, who "invented" the diesel engine, didn't understand their engines either. Mechanical engineering can be like that. At any rate, Stirling, probably in conjunction with his brother, James, did make a working engine, an engine that depends on the behavior of a gas when it is heated and cooled.

To reiterate, the one fact and its corollaries are this: (1) Gas expands when heated and contracts when cooled. (2) Gas absorbs heat when expanded and gives off heat when compressed. (3) The pressure of a gas increases when heated and decreases when cooled. It's worthwhile to consider the other permutations of the one fact: keep volume constant and increase temperature, and pressure increases; increase volume at constant temperature, and pressure drops; decrease volume at constant temperature, and pressure increases; decrease temperature at constant volume, and pressure drops; and so on. This is, in simple terms, the universal gas law, a way of stating the relation between the volume and pressure and temperature of a gas. (The universal gas law also takes into account the mass of the gas, which is what makes it universal.)

A Stirling engine takes advantage of the universal gas law—the ways in which a gas responds to temperature and pressure—by being designed in such a manner as to take energy from heat when there is plenty of heat to be had, using that energy to do work, and then taking advantage of the relative cool that prevails in the cycle because some of the heat has been used to do the work, and because some of the heat has been stored for reuse. A Stirling engine works because it is constructed in such a way that when the gas is doing work on the piston, the gas has a lot of energy with which to do so: it is hot and thus at high pressure; but when the engine is doing work on the gas, the gas is cooler, and thus it takes less energy to do work on the gas.

An engine that could take all of the energy from the gas when the gas was doing the work, and use no energy at all when the engine was doing the work, would be perfect: it would provide perpetual motion, but there are no engines like this. On the other hand, an engine that needed more energy to work on the gas than it got from the gas working on it would be the antithesis of a perpetual-motion device: it would be a lawn ornament, or sculpture, but it would not be an engine. A fair number of engines like this have been made, although it's safe to say that no one ever tried to make one. The best that an engine can do is convert some fraction of the energy put into it into useful work, and it must do this cyclically—a stick of dynamite, for example, isn't really an engine, even though one might use it to do the work of uprooting a tree stump. This is why engines have moving parts—reciprocating, as in a Stirling or an IC engine, or revolving, as in a turbine jet engine: the mechanical parts are configured in such a way that they move in a cycle.

A simple heat engine, then, has a piston fitted into a cylinder. The cylinder is heated at one end, and the gas in the cylinder expands, moving the piston. Just as work expands to fill the time allotted, gas expands to fill the space allotted and will move a piston in order to do so. Now, if the heat source is removed and the gas in the cylinder is *cooled,* say, by dousing the cylinder with cold water, the temperature will drop, as will the pressure, and the piston will move back to somewhere near its original position. Heat a cylinder containing gas, and the pressure and the temperature will rise; douse the cylinder with ice water, and the pressure and temperature will drop. A sealed cylinder with a piston in the middle can be made to do work in just this way: heat one end and cool the other, and drive the piston toward the cold end. To move it back (to make it cyclic), heat the formerly cold end and cool the formerly hot end. Heat. Cool. Heat. Cool. The piston will move back and forth, driven by the expansion of the gas between the piston and the hot end, and complemented by the contraction of the gas between the piston and the cold end. This is what an engineer would call work.

Of course, this would be a hard way to run an engine: apply heat, then remove it and dump cold water on the engine; then heat everything up again. An engine would "run" in this fashion, but the work going in (if one counts the labors of starting a fire, say, to provide heat, and carrying the bucket of cold water, and so on) would be in huge disproportion to the amount of work coming out—it would, in other words, be a very inefficient engine.

A Sunpower Stirling engine, though, rather than having a hot end and a cold end with a piston in between, simply has two different regions, called heat exchangers, where the gas inside it flows; one heat exchanger is kept hot by a heat source, and the other is cooled by water, circulating through tubes or hoses. The temperature difference between the hot place and the cold place is maintained by continuously adding heat to the hot region, removing heat (by circulating water) at the cold region, and connecting the two regions so that the gas in the engine may flow between them. Look at a refrigerator; in relative terms, it has a hot region (the regular refrigerator compartment) and a cold region (the freezer). They are in close proximity to each other, but in one water freezes to ice, and in the other ice melts to water. A cylinder that extended from refrigerator to freezer would be warmer at one end than at the other, although both would feel cold.

Perhaps the simplest way to understand Stirling's engine is to think of *two* cylinders, one kept hot and the other cooled, and with a piston in each connected by a mechanical linkage.

Stirling's engine, in fact, was innovative because of the clever arrangement of linkages: the pistons were connected by the rods and cranks in such a way that when the gas was doing work on the piston, it was doing more work than when the piston was doing work on the gas. It works something like this (the qualifier "something" necessary for reasons to be explained in a moment).

Two cylinders are side by side, connected by a tube, and in each cylinder is a piston. One cylinder is kept hot—Stirling used a coal fire. The other is kept cool—in his first drawings, Stirling had no obvious cooler, or cold sink; his first engines ran, if they ran at all, by using the atmosphere as the cold sink. (In later drawings, a cooler was apparent.) The tube connecting the cylinders is important for two reasons. First, it makes the two cylinders have a common volume—air can move between one cylinder and the other. Second, it contains a material that can absorb heat. Steel wool is good. Almost anything will do, in fact, but steel wool works well and is easy to picture.

The linkage between the pistons is rigged in such a way that the piston in the hot cylinder (call it the "hot piston") starts the cycle halfway in, and the piston in the cold cylinder (the "cold piston") is all the way in. The total volume, then, of the two cylinders and the tube connecting them is relatively small. When the hot piston moves out on its power stroke (pushed by the expanding volume of the heated air in the cylinder), the cold piston remains stationary; because the cold piston stays stationary, the hot, and thus expanding, gas pushes the hot piston out: volume increases, but the temperature in the hot cylinder (because the heat source is continuously adding heat) remains the same, and the pressure in the cylinder *drops:* increase volume at constant temperature, and this will always happen. But the hot piston did move and did, if the pistons can be moved round again to this same position, making a cycle, do work. How, then, to get the piston through a cycle?

Stirling used the linkage connecting the pistons. After the power stroke, the linkage moves the hot piston back in and the cold piston out. This keeps volume constant, but it also forces the hot gas through the tube between the cylinders, through the steel wool. The steel wool absorbs some of the heat from the hot gas and thus cools it—acting, in a way, like the tubes of cool gas in a refrigerator, which absorb heat from the air around them. The steel wool acts as what Stirling people call a "regenerator"—Stirling himself called it an "economiser," because it saved the heat.

So, on this stroke, volume is constant (hot piston in, cold piston out), the temperature of the gas drops (because of the heat absorbed by the steel wool), so pressure drops again: pressure dropped on the first stroke, as volume increased, and now it drops still further, but at constant volume. Now, if the hot piston stays out, and the cold piston moves in, the gas will increase in pressure (because volume is decreasing). But remember, the cold piston is in a cylinder attached to a cold sink, so the temperature doesn't go up: the heat from the rise in pressure is absorbed by the cold sink. So the volume has decreased by a lot, but pressure has increased only slightly. To get back to the position the pistons started out in, the hot piston moves out, and the cold piston moves in. Volume is constant, but the gas moves this time from the cold cylinder to the hot cylinder, passing through the steel wool and *picking up the heat* stored there earlier. This heats the gas and raises the temperature, which raises the pressure and, more important, leaves the hot piston in a position to do work again: the gas is hot and the pressure is high, which is what one wants on the power stroke.

That is the theoretical Stirling cycle. The Stirling cycle of a real working engine (and which, confusingly enough, is often referred to as the *pseudo*-Stirling cycle, to differentiate it from the ideal cycle described above) is slightly different, in large part because the pistons are constantly moving, rather than stopping at the different stages, and because in a real engine there is no such thing as "constant volume," or "constant temperature," or "constant pressure," but rather gradations: no black and white, just gray area.

But the principle that governs both is the core of the cycle: do work when there is plenty of energy to do work with; then use an engineer's tricks—the linkage, the steel wool—to complete the cycle and get ready to do work again.

So what Neill Lane draws on the chalkboard looks like this:

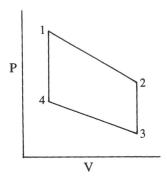

This is a plot that shows the relation between pressure and volume inside an engine at various given stages of a cycle. Up represents an increase in pressure, and to the right represents an increase in volume, as in a cylinder with a piston that can move back and forth. When the piston moves in such a way as to diminish the volume in the cylinder, the pressure (and, remember, the temperature, unless there is a way of taking heat out of the system—a cold sink) goes up; when the piston moves back and increases volume, pressure drops, as does the temperature, unless heat is added to the system.

The figure bounded by the numbers is an indicator diagram of the ideal Stirling cycle. From point 1 to point 2 is the power stroke, showing an increase in volume (because the piston is moving out) and a decrease in pressure; from 2 to 3 is the cooling of the gas by the steel wool (a drop in pressure at constant volume), from 3 to 4 is the decrease in volume with only a slight increase in pressure, and from 4 to 1 is the increase in pressure as the gas takes the heat from the steel wool. It is cyclic. It does work. It is, therefore, an engine.

Why it is an engine, rather than a lawn ornament—or, for that matter, a perpetual-motion machine—is better understood, at least at the beginning, at the level of analogy, rather than that of thermodynamics. Forget, for a moment, heat, and pistons, and steel wool. Think, instead, of *real* wool, like a heavy wool army blanket. Think of Rube Goldberg.

You are in a stairwell, an odd stairwell, but a stairwell all the same. At the top of the stairs stands a tub of water. Next to the flight of stairs is a contraption familiar from county fairs and carnivals, one of those strength-demonstrating devices that look like thermometers. At a carnival, one hits a lever at the bottom of the thermometer with a heavy mallet, and a weight rises to ring a bell. If you're strong enough, that is. This one operates only slightly differently. A wool blanket soaked with water is heavy enough, if dropped from the top of the stairs onto the lever, to make the weight rise and ring the bell. When the bell rings on this device, a slot opens and a ten-dollar bill flutters out, money to pay the rent and upkeep on the stairwell and the carnival contraption, with some left over to spend on whatever you want, if you can make the bell ring often enough.

Just as at the carnival though, it takes a certain amount of power to make the bell ring. A dry blanket dropped from the top of the stairs won't ring the bell; only the soaking-wet blanket will. But the stairs are steep. You might soak the blanket in water, drop it over the ledge, collect your ten dollars, and try hauling the wet blanket up the stairs. After only one or two trips, though, you're tired. What you need is something that won't keep the blanket from being heavy enough to ring the bell, but that will make the blanket easier to carry upstairs. You buy a clothes dryer and install it at the bottom of the stairs.

Now the trick to the Stirling engine seems obvious. Drop the wet blanket over the edge and collect the ten dollars, but dry the blanket in the clothes dryer until it is fluffy and light before you carry it up the stairs. Soak it in the tub, drop it over the edge, collect the ten dollars, and dry the blanket before you carry it back up. The clothes dryer adds some expense to the overhead, but you still make a profit, because you're able to make many more trips. There are many small costs associated with this way of making money—water to replenish the tub from time to time, electricity for the clothes dryer, and so on, but if you make even a tiny profit, you're ahead of the game. You (and your stairwell contraption) are doing what an engineer would call work.

Broadly scrawled, the analogy works this way. The water is heat and the blanket the working fluid. The clothes dryer doesn't fit perfectly in place of the steel wool—to be a closer fit, it would have to have a way of getting the water out of the blanket and back into the tank via, say, a condenser at the top of the stairwell into which one runs the flexible white tube that comes out of the back of a clothes dryer, to recycle the water taken from the blanket. The ten-dollar bills can be seen as the gross work of the cycle, and what's left over after paying for the electricity, the water, the blanket, the rent, and the wear and tear is the net work. This analogy even fits, after a fashion, within an indicator diagram: from 1 to 2 is the dropping of the blanket onto the lever—the power stroke, where the water in the blanket makes the blanket heavy enough to move the lever; it acts like an expanding gas in a cylinder, moving a piston. Moving the wet blanket from the lever to the dryer is close to the move from 2 to 3; moving the light, fluffy blanket up the stairs is that from 3 to 4, and soaking the blanket in the water caught from the dryer vent is equivalent to that from 4 to 1. It's not a perfect analogy—and thermodynamicists will *hate* it—but it's not bad.

The whole contraption in the stairwell is, however, functional according to the same principles operative in a Stirling engine: it uses energy when there is plenty of energy around to get the work out of the system, and then takes advantage of the conditions that prevail after the work is done in order to get back to the beginning of the cycle. As with any engine, once the principles are understood, the opportunities for improvement, for streamlining, are obvious. The first generation of any invention is bound to seem, many generations later, what people at Sunpower call kludgy—clumsy, contrived, ugly. It's interesting to note that Otto—he of the internal-combustion engine—would have found the stairwell Stirling very familiar: the first Otto engine had long rods on which weights slid up and down, powered by explosions of fuel at the bottom; the falling weights hit a ratchet arrangement before being blown back up the rod. Otto, indeed, would probably have found it too familiar for comfort. However, once one figures out how to get work out of an engine, the rest is refinement. And so the Stirling engine existed for more than 150 years, before William Beale saw in it something that made all pre-Beale engines look kludgy. Thousands of engineers had looked at the Stirling, had worked with it, refined it, changed it, tinkered with it. William saw a way to remake it into the perfect engine for the twenty-first century.

Chapter 3

The Ultimate Machine

Quiet, reliable, only two moving parts: Stirling engines last forever.

—*early Sunpower advertising slogan*

I've never been inhibited by things I can't analyze.

—*one version of a famous William Beale comment during a technical meeting*

My inability to analyze a problem has never interfered with my attempts to solve it.

—*the other version*

QUIET. RELIABLE. TWO MOVING PARTS. LAST FOREVER. This is one of Sunpower's original advertising slogans, and it is hyperbolic in the way most advertising slogans are. In theory, it is a schematic of the Beale free-piston Stirling engine, just as in theory, say, a certain brand of automobile will improve your social standing or a certain kind of soap will make your life sweeter. In point of fact, the addition of the words "in theory" to the slogan makes it absolutely true.

The way William Beale tells it is simple; he was teaching a mechanical-engineering class at Ohio University, where he was a professor, and he assigned a Stirling-engine problem to the ME students. In doing so, he showed them drawings of a fairly typical Stirling design, one with a linkage system similar to Stirling's own invention, in which the movement of the pistons was governed by a crankshaft and rod arrangement. He also had a sudden flash of recognition: "Cut the cranks off a Stirling engine, and it will *still* work," was what he said at that moment, and the free-piston Stirling engine, or at least the idea of it, was born.

To understand the many layers of significance in this flash of recognition requires some peeling away, as with an onion. What William Beale saw led him to the idea of Sunpower, led him to the vision of a world powered by Stirling engines, led him to imagine a world with many fewer problems. It turned him from an associate professor of mechanical engineering into a visionary. Visionaries can make such leaps; that's why they're visionaries. The rest of us go step by step.

Why is a Stirling engine with its cranks cut off a machine that can change the world? The answer is in two parts. The first is thermodynamic; the second, mechanical. The thermodynamic aspects of the Stirling cycle are well known and well defined; the practical mechanical limitations were also thought to be well understood, until William Beale thought to take a hacksaw (metaphorically) to a Stirling engine. The idea came to him suddenly; the application has occupied the last quarter century.

That William Beale—or any associate professor of ME, for that matter—was familiar with the Stirling engine at all has to do with the thermodynamics of heat engines. It has to do with the second law, and with a French mathematician named Carnot. As befits a history where Stirling perhaps didn't understand his own engine, where Otto perhaps didn't understand his own engine, where Diesel perhaps didn't understand *his* own engine, Carnot most certainly didn't understand important aspects of his own work. In this sense, then, he makes a good Father of Heat Engines. He knew—somehow—how something could be, without knowing exactly *why* it was so: his inability to analyze a problem, in other words, didn't seem to inhibit his attempts to solve it. This is not to say that William Beale is another Carnot; rather, it is to say perhaps, as Louis Pasteur did in another context, that chance favors the prepared mind. Carnot based his understanding and theoretical interpretation of thermodynamics on a flawed premise, but the *result* turned out to be profoundly correct.

Nicolas-Léonard-Sadi Carnot (1796–1832) deserves a book of his own; he fought in Napoleon's army in the defense of Paris in 1814, an experience that, after the war, led him to his analysis of engines and mechanical power: P. W. Atkins sums up Carnot's vision as seeing the steam engine (an important part of the British armaments industry and thus crucial to Britain's victory at Waterloo in 1815) as a device that would "enlarge humanity's social and economic horizons, and carry it into a new world of achievement." Carnot sought to analyze the device that he saw would change the world. That his theories were based on the existence of a nonexistent substance known as "caloric"—a massless fluid of pure heat—did not prevent him from making judgments about the intrinsic losses when heat is converted into work, judgments that along with the work of those who followed him evolved into what we call the laws of thermodynamics.

Carnot's theoretical "engines" depended on caloric to run; that there is no such substance as caloric doesn't mean that Carnot was wrong, exactly. It can be taken as meaning that he could see beyond the limitations of science in the time that he lived, because heat and energy behave in much the way that Carnot envisioned caloric as behaving. He had an inkling of the relation between heat and work, but not the vocabulary to explain it.

Carnot is pertinent to William Beale, and the free-piston Stirling engine, in this way. The "Carnot cycle" defines how an engine *could* work; it can be described by an indicator diagram, just as Stirling's cycle can, although the Carnot engine is for all practical purposes only theoretical, because the amount of work that would be produced is infinitesimally small. The Carnot cycle is important because it depends on the existence not only of a heat source—which all engines, and engine theories, of the time did—but also of a cold sink. It was Carnot who presaged the thermodynamic basis for engines by describing a relationship between the energy going in (in the form of heat), the work coming out, and *the energy lost, irretrievably,* in the process. It was Carnot who established the importance of the "cold sink" in heat engines, as the place where the heat that is not converted into work must go. From Carnot comes our understanding of the cost of doing business, and of the idea of entropy—that there is a loss in all attempts to convert heat into work.

To put it slightly differently, from Carnot comes the recognition that all of the heat put into an engine cannot be converted to work— that no perpetual-motion machine is possible, but also that the efficiency of an engine can be described as being some percentage of the energy put in that comes out as work. There is more to Carnot's work than this, of course, as well as to the work of a dozen other scientists who refined and defined the field of thermodynamics—Kelvin, Joule, Clausius, Boltzmann—but it is from Carnot that a governing relationship of mechanical engineering is drawn: put simply, it is that the efficiency of an engine in converting heat into work is a function of the relationship between the temperature of the heat source and the temperature of the cold sink. Students in a thermodynamics class learn

$$1 - (\text{Temperature}_{\text{cold sink}} / \text{Temperature}_{\text{heat source}})$$

almost before they learn anything else: they learn that the defining characteristic of theoretical efficiency is independent of everything but the temperature of the heat source of the engine and that of the cold sink of the engine. All else is subordinate to this theoretical efficiency.

It is important to say "theoretical" here in that all manner of factors can rob an engine of *practical* efficiency—mechanical losses such as friction—but in an imaginary engine that is free of all mechanical losses, the absolute efficiency is determined by the ratio of the cold-sink temperature to the heat source temperature.

Another way to look at it is to look at the area that is encompassed by the indicator diagram that Neill Lane draws on the chalkboard; the area inside the indicator diagram is a schematic of the work that an engine can do—its theoretical efficiency. The larger the area inside, the higher the theoretical potential. A *real* Stirling engine, however, has a smaller, as well as rounder, indicator diagram. It is smaller for the same reason it is more round; there are no absolute changes—*bang!*—from high pressure to low, or high temperature to low, in the real world. In the real world, there is no practical way (except in an experiment specifically designed to attain it) to have a large increase in pressure at constant temperature. So the real indicator diagram for a real Stirling engine looks different from the impossible ideal; the indicator diagram of a real Stirling engine fits neatly inside that of the ideal, like this:

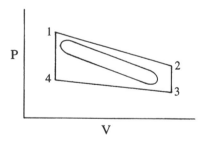

The banana shape is the indicator diagram that a real Stirling engine might produce; the straight lines are the theoretical ideal.

The ideal Stirling cycle offers a very high theoretical efficiency: its theoretical efficiency is in fact that of the Carnot cycle. An ideal Stirling cycle engine can't be built, but a practical, or "pseudo," Stirling engine *can*. So the Stirling cycle is taught in engineering schools, because all the engineering textbooks say (in their one or two paragraphs devoted to the Stirling cycle) that the Stirling represents a very high theoretical efficiency for an engine. Students draw the ideal cycle as an indicator

diagram and see the potential—the theoretical potential—for high efficiency.

But theoretical efficiency, alas, is like imagining money: it is easy
to daydream of a billion dollars, but a hundred-dollar bill, though much
more modest than imagined riches, is actually *worth* something. And
the Stirling engine's history and reputation was clear: terrific theoretical
efficiency. Poor practical efficiency. And it is only practical efficiency
that one can take to the bank.

To understand why Stirlings were so impractical in terms of efficiency is to understand the art of mechanical engineering, to make the
move from theoretical to real: mechanical engineering is a *tangible* field.
The original Stirling engines, remember, had two noteworthy features:
they operated at or near atmospheric pressure (a virtue in a time when
engines at higher pressures were prone to exploding), and they operated at all because of Stirling's clever crankshaft and linkage between
the pistons. Materials science took care of the explosions and permitted
virtually limitless operating pressures; nothing, until William Beale,
could eliminate the need for a crankshaft.

All Stirling engines prior to William Beale had some sort of rod or
shaft or crank apparatus; the rods and cranks connected the pistons,
whether they rode in separate cylinders or in the same cylinder. This
not only made sense; it was crucial to making a Stirling engine work,
because the crank apparatus was what moved the pistons in the cycle of
increasing and decreasing volumes necessary for the engine to work.
But the most obvious need for a shaft of some kind was as simple as it
was obvious: it was the method by which the power of the moving
piston got from inside the engine to the outside world, where the
power, in the form of mechanical motion, could be used.

The most significant change in the technology until William Beale
was subtle, but necessary; rather than two cylinders side by side connected by a tube, Stirling also made an engine that combined the two
cylinders into one, with one piston atop the other, its rod passing
through the center of the lower piston. This not only consolidated the
structure handsomely; it permitted more-efficient heat exchange.

As the technology evolved, the engines got smaller and more efficient, although the early engines, running at or near atmospheric pressure, turned out to be no match for internal-combustion engines for
many nineteenth- and early-twentieth-century uses; their efficiency was
limited by the very fact that made the Stirling a worthwhile invention in
the first place. If you increase the pressure at which the engine operates,
several things happen, but the most important in terms of efficiency is

simple: if you double the pressure, you double the density of the gas inside the engine, which can then transfer double the amount of heat (in round numbers, and ignoring some of the characteristics of gases under pressure for the time being). The higher the pressure, then, the more power produced by the engine, because heat is such an important characteristic of what determines an engine's efficiency.

But Stirling engines for their first 150 years were all based on Stirling's original design, which had the pistons connected by rods that turned cranks. To run them at higher pressures required sophisticated seals and gaskets, designed to let the power (via a rod attached to a crankshaft) out but to keep the gas, under pressure, in.

And a moving rod under high pressure can be the devil to seal, as generation after generation of Stirling-machine designers found out. The seals leaked; they wore out; the ones that worked best had to be made out of exotic and expensive materials or manufactured with exquisite precision. As recently as the 1970s, research programs looking for a way to use a Stirling engine to run an automobile foundered, again and again, on the seal problem. During periods of energy awareness (or expensive fuel, such as during the Arab oil embargo of the early 1970s), Stirling R & D programs arose, with whole teams of engineers working on seals.

But the seal problem has never been solved, really: it seemed that the best the Stirling could offer was a relatively low-pressure engine, which meant large for its power. So Stirling engines all but disappeared, although they were resurrected from time to time for certain specialized uses; some powered oceangoing ships and ferries; others ran small appliances like circulating fans. Still others were constructed on a truly colossal scale, but not in any great numbers. It wasn't until William Beale, in fact, that the Stirling engine of the late twentieth century would look any different from the Stirling engine of the early nineteenth. To say that William Beale solved the seal problem is not quite accurate: he simply eliminated the need for the seals entirely.

Why this was so is easy to visualize. Picture a Stirling engine as having a single cylinder, one piston atop the other. Call the top piston the displacer, because its main job is not to provide the power stroke of the engine, but rather, to take up space: a good displacer is shaped like a can, and its volume forces the gas in the engine through the heat exchangers. The lower piston is the power piston; instead of compressing the gas against the sealed end of the cylinder, it compresses the gas against the bottom of the other piston. (An ice cream treat called the Push-Up comes to mind here, with the displacer being analogous to the ice cream and the cardboard disk attached to the stick analogous to the power piston. Push the ice cream up, and then pull the stick back, and you have a cylinder with two pistons, just like a Stirling engine.)

The pistons still seem to need some form of linkage between them to make them move in the proper sequence, just as when they were side by side in dual cylinders, but when the engine is running, there is a way of seeing the engine that is topsy-turvy: Yeats asked, "How to tell the dancer from the dance?" Beale asked, "Is the linkage moving the pistons, or are the pistons moving the linkage?"

Without the linkage—cut the crank *off,* and remove the linkages that connect them—the pistons will still move: heat pours in from the heat source and expands the gas, and the pistons move. But in a sealed vessel, a cylinder where the bottom piston—the power piston—has a blank wall beneath it, the piston is being pushed by the expanding gas on one side, but at the same time is *compressing* gas on the other. Squeeze a balloon, and the balloon squeezes back. So the piston "bounces" back up, compressing the hot gas, forcing it through the steel wool, and so on. An indicator diagram of the cycle works, and shows work being done: heat pours in, the piston moves out; pressure drops at the hot end of the cylinder at the same time it's rising beneath it, which forces the displacer up and pushes the gas into the cool region—the cold sink. The piston is "bounced" back up by the higher pressure at the closed end of the cylinder, which also moves gas and forces the displacer away. It's an odd sort of dance, but it works. In a sealed cylinder with a heat source and cold sink connected by a tube containing the steel wool, the pistons move up and back, expanding, bouncing, compressing, and displacing. The engine works, and the engine does work. This, schematically, is a free-piston Stirling engine, an FPSE. The piston and the displacer are not connected to each other, or to anything else, but they move back and forth.

The potential for operating the pistons inside a sealed cylinder eliminates the rod and seal problem, and thus makes operating pressure a variable that the engine designer can manipulate, because there are no seals to worry about. (It also, of course, raises the question of how to get the work out of the engine, a different, but also solvable, problem.)

Stirling's engines ran near atmospheric pressure, which is about 15 pounds per square inch. Sunpower has made engines that run at pressures as high as 1,500 pounds per square inch (although, it should be noted, no one really likes to work with pressures this high); the 5K has an operating pressure of about 650 pounds per square inch, which means that some forty times as much gas can be inside the engine and carrying heat than in Stirling's Stirling. This alone opens up a whole new range of possibilities.

Another range of possibilities is suggested by Sunpower's advertising slogan. Stirling engines *are* quiet, no question about that, and in the simplest FPSE there could be just two moving parts: a piston and a displacer. Such an engine could be made to last, if not forever, for a very long time, although there is always a tradeoff involved: to get long life, something else is given up—power or efficiency. When engineers begin to trade up in favor of efficiency, or down in favor of size, certain factors come into play that can make Stirling engines lose life expectancy. William Beale and the engineers at Sunpower have plenty of ideas about this, of course—one of William's most oft-stated facts is that "well, we've solved the life problem." This is also one of the most oft-*quoted* facts around Sunpower, said generally just after something has broken down.

But building an engine that runs forever is a luxurious consideration in many ways; engines need to run in the first place before they can run for eternity. And while the FPSE might seem to solve one problem, it also seems to cause another, at least to the layperson: how to get the power *out?* To a mechanical engineer, this is trivial, at least at first glance; engineers immediately envision pumps when they see a reciprocating device, because many pumps operate in this manner, as anyone who has operated a hand pump on a farm or at a park or campground knows: a piston in a cylinder is drawn up, and the vacuum beneath it draws water up with it. That most of the contemporary world operates on rotating power devices is of no theoretical significance, necessarily: most engines produce rotational power and thus beget a world of work where things go round and round: crankshafts, motors, and the like. Reciprocating motion, though, is good for at least one other thing, besides pumping water. It's ideal for making the *alternating* current used as electricity. The FPSE seems to tie together, quite neatly, a handful of apparently unconnected facts into an engine of change.

The history of science—indeed, of all human endeavor—tends to smooth out the edges, just as the indicator diagram of a real Stirling engine smooths out the square corners of the ideal. The hindsight we use to understand change is a peculiar lens, arranging facts and events in an order that seems to lead, irrevocably, to the present. To look at the Stirling engine this way is myopic, then, but provides a simplified, a compressed, time line from Stirling to Beale. It goes like this: Stirling invented an engine with high theoretical efficiency, but low practical efficiency; Beale refined Stirling's engine in such a way as to improve the practical efficiency to be a much larger fraction of theoretical efficiency than was previously thought possible, and he did so by means of an engine that, efficiency aside, is highly adaptable and highly versatile, especially in terms of fuel. This versatility is one of the two most important changes wrought by the FPSE, in that it raises questions that before had been ignored largely because there was no technology to answer them.

We live, at the moment, in an internal-combustion-powered world, in large part because there has been no alternative. The FPSE conjures up an image of a Stirling-powered world, because it has a potentially satisfactory answer to some of those questions: the answer borders on the *smug,* in fact, because a solar-powered Stirling changes profoundly the thermodynamic equations our world operates within. The FPSE does not *change* the second law (nothing can), but it does suggest a different set of definitions.

A century ago, technology was considered somewhat differently from how it is viewed today: the mature technologies of the time, like the steam engine, were rapidly giving way to newer adaptations, like the internal-combustion engine, in the way that all technologies are eventually replaced: invention, refinement, adaptation, adoption, development. It is a synthetic process, first and foremost. An internal-combustion engine, for example is useless without a reliable source of fuel; without the discovery of oil, and the refinement of the refining process to provide a steady and plentiful supply, the internal-combustion engine would have foundered. Instead of foundering, though, the IC engine thrived: the economic, political, cultural, and social climate was perfect for the IC engine, and so an infrastructure was created where none had been before, and we are learning to live with the consequences.

The costs, both direct and indirect, associated with the development and use of any technology can be calculated, if one takes the trouble to do so: it is possible to determine, for example, just how many tons of carbon dioxide, of sulfur and nitrogen, of unburned hydrocarbons, automobiles put into the atmosphere. Likewise, electric power plants that burn coal know how much carbon dioxide they take out of the ground and disperse into the air. All of this is known, because we have become more and more conscious of the significance of such activity.

Nowadays, though, *no* technology comes to market without a thorough examination and evaluation of its costs and benefits, which is something the world has learned from the technology it now possesses. But this has been learned the hard way: no one thought to ask whether the automobile, for example, entailed long-term costs to the environment that should have been taken into consideration when the automobile was evaluated as a product of technology. Now it is too late to do so: the automobile is with us to stay. To eliminate the automobile is unimaginable, so instead it is modified, tinkered with, adjusted, to get more miles per gallon, to emit fewer parts per million of carbon dioxide, to cause less violent an injury when it crashes into another automobile.

This is also the case, although much less apparently, with the way industrialized countries generate the energy they use: one of the first fuels ever used for energy production was coal, and millions of tons of coal are still used today. We may burn it more efficiently, we may "scrub" the emissions with sophisticated and complicated machines, we may burn "clean" coal, but we still burn coal, coal that has lain underground for millions of years. The same is so for oil. The infrastructure, like the infrastructure that supports the automobile, seems firmly in place. This is most obvious in the industrialized world, although we are as rapidly as possible exporting to developing countries all of the technologies, all of the infrastructures, that we are now finding to be less and less visionary, and more and more troublesome.

The free-piston Stirling engine—even the *idea* of a free-piston Stirling engine—suggests a very different world. Students of the second law put it this way: right now, we are moving stored energy through time, by extracting resources from the ground that decayed there millions of years ago, and burning them—releasing the energy—in our own time. Solar energy doesn't do this; there is still a movement of energy, but it is through space; not through millions of years, but over a few thousand miles. Take heat from the Arizona desert and move it, in the form of electricity, to New York City, and release it there. This is the second law: you can't get rid of heat, you can only move it around. But we can move it through time, as when we mine coal or pump oil, or we can move it through space, as when we take heat from the sun in one place and release it somewhere else. Given that you can't *beat* the second law, what do you do?

With fossil fuels, you take heat from eons ago and release it today, because there is no real alternative. The FPSE hints at an alternative. Call it spatial displacement, rather than time displacement, of heat. It is a heady concept. No one knows exactly what would happen if we changed from time displacement to spatial displacement: William Beale believes, for example, that the Arizona desert would become more temperate—take one-third of the heat out of the desert (which is about what a good solar collector will gather), and the desert cools off. And since the heat would be so widely dispersed—lots of heat, to be sure, but scattered everywhere in small amounts—the net effect wouldn't seem to be bad. But things would be *different*. We would no longer be increasing the net amount of carbon dioxide in our atmosphere; we wouldn't be making acid rain. Automobiles would run on hydrogen, which is a very clean fuel (the "exhaust" is water vapor) that's easy to make with electricity. The world's energy problems seem to evaporate around Sunpower, on good days at least, because the FPSE seems to fit so well with our current crisis. That's the theory, anyway. The FPSE, as refined by Beale, ties nicely into a theoretical framework that would ultimately make fossil fuels obsolete. All you have to do is make the machine, which is what they are trying to do at Sunpower, which is what, arguably, the dozen or so engineers at Sunpower are the most experienced in the world at doing.

It turns out, though, that *(a)* making a Stirling engine is easy, but *(b)* making a Stirling engine that can compete on a playing field skewed toward fossil fuels is much harder, which means that *(c)* one has to pull out all the stops and *(d)* do it with not very much money, support, or understanding; *(e)*, which is telling the world how misguided it is, is sometimes satisfying, but not very productive. Otto got away with a Rube Goldbergish device for the first IC engine, because he was competing against technologies that had many strikes against them; Sunpower has the unenviable task of competing against technologies that have an infrastructure so intertwined with our daily lives that we can hardly imagine a world without them. Neill Lane likes to offer an analogy to illustrate this. "Imagine," he says, "that we had the world that we have, identical in every way, except that there was no electrical grid—no utility poles, no high-tension wires, no wires running from the pole on the corner to each house on the block. Now. Where would you start, if you wanted to install a web of wires and transmission stations and cables to every house, every office building, every factory? It's incalculable; you'd go to Washington and say, 'Well, we need a trillion dollars worth of copper, and several hundred billions of barrels of oil to make the cable insulation, and, oh, several hundred million poles to string the wires on, and, oh yes, a labor force of tens of millions to install everything, and it's going to take us fifty years.' The grid is here to stay, no matter *what* gets invented, because it would take the GNP for a couple of decades to replace it."

So Sunpower engineers can't simply make a "better" Stirling engine. They have to make an *appropriate* one, one that fits in with the established infrastructure. Otherwise, they are only tinkering with theory.

But "better," it seems unnecessary to state, doesn't carry with it automatic acceptance, any more than complexity means a device will be shunned (look under the hood of a new automobile), or danger means a technology will falter (look under the hood of a nuclear power plant). Cultures, societies, choose the devices that will be ascendant according to their own immediate needs and what they find available with which to work, and societies change their technology only when reasons for change become too compelling to be ignored, not because things are better.

"Better," too, is a slippery word; in some contexts, better might mean burning fossil fuels as fast as we can dig them up or conquer countries that have them; in a very short-term economic sense, it's hard to describe a better fuel than gasoline, which is cheaper in real dollars today than it was in the 1960s. It's only when all the costs are considered—to health, environment, policy, society—that gasoline looks expensive. In Sunpower dollars, which might be described as *total* cost per kilowatt hour, the Stirling looks impossibly good, but only if that "total" includes the cost of, say, acid rain in Canada, or a war in the Persian Gulf, or the future cost of building dikes around Manhattan when the ice caps begin to melt. On a very simple level, solar power looks as though it ought to be cheap—if a coal-fired power plant and a number of solar Stirlings cost about the same to build and produce the same number of kilowatts, the Stirling seems to have an edge, if only because its fuel, solar power, is free. In the past, solar power was stunted because previous solar technology was cost-effective only when providing power to places in or near the desert, or in tiny amounts at negligible voltages. The FPSE would make electricity that ties into the existing grid and thus would be as portable as electricity can be, which is highly portable—we already ship electricity thousands of miles, as a matter of course.

But a solar FPSE can't just be better; it has to be what they call around Sunpower an "ultimate machine." It has to be quiet and reliable; it has to last, if not forever, at least for a very long time. It should have as few moving parts as possible, although the days of the two-moving-part FPSE are long gone, if they ever existed at all (two *big* moving parts, okay; but there are some small ones). The ultimate machine will need to be all of this, and more. It needs, too, to fit into the real world, though, because it will have to be the real world that funds its development and that buys the finished product.

It's not impossible. All Stirling engines work on the chalkboard or a sketch pad; most work when simulated by means of a computer program; a fair number of Stirling engines that make it from the drawing board into hardware work, in one fashion or another. Some few work badly and, occasionally, not at all, which is what can happen in R & D. It was not long ago that a working Stirling engine at Sunpower was cause for earnest celebration; not much before that, it caused stupefaction. David Berchowitz, who came to Sunpower in 1981, remembers when his first Sunpower Stirling engine ran on the first try. For about twenty seconds. "William called me into his office," says David, "and I thought sure it was to *fire* me. Instead, he gave me a *raise*." One engine at Sunpower is legendary for having put out about 90 percent of its design power in its first runs, not a bad start at all. After a year's worth of very intense work, it still put out 90 percent of its design power, which is not much of a finish. Invention in any field is like this; invention in the Stirling engine field is more like this than like anything else. Invention, research, and development are a long way from the stage that technology aspires to, a stage when technology may be taken for granted. Indeed, Sunpower's first-run parties are reminders of a time, not too long ago, when the switch was thrown to try to start an engine, and everyone—designer, machinist, technician—gathered round to see what might happen.

What works in theory, then, must be made to work in hardware, and must be made to work in hardware in such a way that it's not grounds for a party when something *does* work. The two leaps that have to be made on every engine are always the same: one is performance and the other is reliability. Only when Sunpower made a machine that did what it was supposed to, and did it for a reasonable length of time, would an ultimate machine exist.

It goes without saying that everyone at Sunpower has very good ideas as to how this should be accomplished—and that most people there are not shy about sharing those ideas. This is a large part of Sunpower's history, a seat-of-the-pants sort of place where amazing things could happen. Often, those amazing things were good. Occasionally, they were baffling. All too often, just when the amazing things started to happen, the money, or the time, or the sponsor, would run out, and it would be back to square one. Different sponsor, different requirements, constraints, criteria. Throughout its history, Sunpower has found itself reinventing, redesigning, refabricating, and so on, for virtually every engine, throwing out babies with bathwater because the old baby didn't appeal to the new sponsor.

A good example of this is the small engine, one of William's most long-lived dreams. In the universe according to William, the small engine would be compact, simple, adaptable to many different fuels and uses: the same engine would form the heart of a home cogeneration system (the box in the basement) and a remote solar-powered generator. It would have an optional burner for coal or wood chips or cow dung or rice husks, and a burner for natural gas or propane. It would be rugged enough for a Mongolian goatherd to haul around when he moves his yurt, yet light enough to strap to the back of a horse. Small enough to be unobtrusive at a campsite or on a pleasure boat, but powerful enough to displace the need for a fossil-fuel-powered generator. All of these are, to greater and lesser degrees, conflicting goals. An engine that runs off a solar dish the size of a small television satellite dish, for instance, *needs* a certain mass, because it would be balanced against the mass of the dish itself. An engine that runs on propane runs best with a certain sort of heat exchanger; an engine that uses cow manure runs best with an entirely different heat exchanger. Should it make alternating current, so as to be compatible with widely used electric devices, or should it make direct current, so that it can charge a battery? Should it be modified to make both? William's ultimate machine would be adaptable to all of these things: customers would pick out items on a list to make the system that most closely fit their needs. But a sponsor who wants to sell dung-burning engines to Mongolia doesn't want to fund the development of a natural-gas burner for use in an American home. Likewise, a sponsor interested in selling Stirling generators to motor home owners couldn't care less about the problems faced by a Mongolian goatherd. So not only does Sunpower find itself in the position of having to, almost literally, reinvent the universe; it has to invent it one discrete chunk at a time. When it can adapt or convert, it does so, but it's not the way a company with a billion dollars would develop an ultimate machine. It's the way a company with very small change gets by.

That Sunpower operates on very small change is probably not surprising—most small companies do. That it has operated on a shoestring for a quarter of a century is perhaps more surprising, although not unheard of. That it exists today in much the same form, with the same goals and ideals, with exactly the same mission, might raise some eyebrows. That several millions of dollars, dozens of engineers, a handful of spinoffs and subsidiaries and diversifications, and literally hundreds of thousands of man-hours have been consumed by a company that still doesn't have something to sell just yet may start to sound more than a bit strange.

That one man conceived the idea, the company, the business, and the market, and that one man has been in charge and, furthermore, has outlasted financial crisis, technical catastrophe, bad luck, and five American presidents, all the while searching doggedly for a Holy Grail that he knows is out there but most people don't care about, and who *still* gets up every morning and goes to work in order to change the way the world makes energy—well, if William Beale cannot be described as the most determined human being on the planet, it is because the word "determined" doesn't go far enough.

One rainy afternoon at Sunpower, two engineers sat, legs sprawled out like punch-drunk boxers, in the drafting room. The rest of the building was dark, almost everyone else having gone home. The two engineers looked miserable—they *were* miserable—because the company was out of money, the two projects on which the engineers worked were out of money, they were both facing hugely problematic setbacks that even with lots of money would be nerve-racking to face. Engineers, when they are down and out, tell the most depressing stories they know: about explosions, fires, collapsed bridges, humiliating failure, abject loss. Sometimes they do it because it puts their own engineering tasks in perspective; sometimes they do it because engineers are obsessed with making things work and understanding why certain things don't. Sometimes such conversations are cathartic; at other times, they are just sort of depressing.

Then they started telling William Beale stories, some of which were from their own experience at Sunpower and others from Sunpower lore and legend, twenty-plus years of history centered on one gruff, rumpled, shaggy mechanical engineer whom they called Bwana William. Plenty of these stories were about engineering failure, too, the sort of failure that is endemic to research and invention, to trying to do something that hasn't been done before. Then they start talking about Bwana William himself, before the conversation finally peters out, and they sit silent for a while, gathering the energy to go down to the West End Tavern to do product testing on yeast beverages and R & D on malt.

"Do you think William Beale is unique?" one engineer asked the other idly, musing almost to himself.

After a pause, the other engineer answered, *"Jesus God, I hope so."*

Chapter **4**

The World according to William

As a result of the "Energy Crisis," the world has rather belatedly come to recognize an obvious fact—that many present sources of energy are limited and that the present methods of use are extremely wasteful. . . .

—*William Beale, in* Sunpower Newsletter, *number 1, January 2, 1974*

A SORT OF JAUNTINESS still inhabits sixty-five-year-old William T. Beale, against all odds: to have worked at Sunpower from the beginning has been to toil amid daily frustrations, frustrations that stretch out over weeks before becoming disappointments, disappointments that then turn, over months, into seemingly hopeless tasks, tasks that metamorphose, over years, into apparently fruitless endeavors, endeavors that, when they begin to span decades, can look impossibly flawed. When one lives a life this way, jauntiness is generally the first thing to go.

And jauntiness is absent for stretches even in William, replaced temporarily by grief, or anger, or irritation, until his enthusiasms are again fired by some small success, some bright idea, some moment of thermodynamic delicacy, and he is again off on the hunt, two and a half decades of thankless effort and penurious past behind him in an instant as he reexperiences, relives, the magic and beauty of an idea that *works*. For it is not the rock against which William pushes that keeps him from changing the world, but rather the chains of an unimaginative culture that he has not yet been able to kick away from his ankles. Part Sisyphus, part Prometheus, but jaunty to the last.

During the dark days of the summer of 1991, when Sunpower

money is not so much tight as nonexistent, William is supposed to be chained to his desk, working the phone, beating the bushes, drumming up trade, stirring the waters, all Sunpower expressions for nonengineering efforts. But where he wants to be is in the lab.

Such expressions for trying to conjure up financial support for the company are ubiquitous at Sunpower, as befits an activity that goes on, at crisis pitch, most of the time. And, since it is a company first and foremost of engineers, phrases that sound like clichés to those in the outside world seem instead to be gems of substance around Sunpower, expressions that, with economy and vividness, describe some other-worldly activity, which is anything that isn't engineering activity. "Talking" at Sunpower, then, is "gabbing" or "gassing" or "shooting the breeze"; it is done on the "horn," with "desk jockeys" and "pencil pushers" and "bean counters" and "money men"—"men in suits," in other words—at every place other than Sunpower. *Real* engineers only behave like paper pushers, they only gas and gab and shoot the breeze on the horn, when they are "chained to the desk."

For hours at a stretch during such crisis times, William indeed manages to stay in harness, to be a desk jockey, to push paper, to play phone tag with men in suits, to gab, to gas, to blow hot air, to get on the horn and *stay* on the horn in order to beat the bushes. "I'm not supposed to be out here," he says one morning in the lab to Chuck Howenstine. "Crawford has me chained to the desk."

Chuck is fabricating a new solid-fuel burner for the small engine, a project so *notoriously* unfunded that it has become a running joke at Sunpower, the project that wouldn't die. And for weeks at a time, the various components of the small engine lie untouched, until temporary backups or fallow periods on funded projects free Chuck to spend an hour or so at his table in the lab. When William makes his rounds ("just stretching my legs," he says), he saunters through the big swinging doors into the lab, hands in his pockets, a man just out for a stroll before getting back to his desk. William pokes around at Dave Shade's table, peeks into the heat pump test cell, picks up a tool or a part from Al Schubert's workbench and sets it down again, and then casually stops at Chuck's workbench and watches Chuck work.

Chuck Howenstine is exactly the sort of person to work well with William. Chuck is patient and methodical (in direct contrast with William), meticulous out of sheer habit (a large, neatly carved sign over Chuck's workbench says, "Know the workman by his work"), and inquisitive for its own sake. Best of all, Chuck is, as William is, a tinkerer to the *core,* the sort of tinkerer who built a Stirling engine years before he had ever even heard of Sunpower, out of nothing more than innate curiosity. A champion at scrounging useful items from the trash (his Stirling engine contained not a *single* store-bought component), Chuck has his heyday every spring when the students of Ohio University clean out their rented tenements and leave town. "You wouldn't *believe* the good stuff some people will throw away," Chuck will say, displaying some bit of good stuff he found on his bicycle ride home. Chuck is thrifty and clever and sincere, the sort of person who would never use a new drill bit when he can sharpen an old one, who never throws away scratch paper until both sides are covered with sketches or calculations, who always turns off the lights when he leaves a room. Chuck lives an economical, almost Thoreauvian, life: economy as poetry.

Which makes him a perfect technician for the small engine, in that the small engine not only has no current income but has a rather substantial internal debt, having consumed a significant fraction of Sunpower's R & D money over the course of its life, money that was earmarked, in part, for other projects. In the Monday morning meeting minutes (and in the meeting itself, for that matter), even the small engine's *name* has been trimmed: Peggy Shank and Elaine Mather refer to the program simply as "the small." But the small engine is, as a concept, a philosophy, and a metaphor, William's baby, and hence it has yet to be thrown out with the bathwater, although there have been moments, days when all the beating of the bushes produces nothing but sore arms, when the small looked dead.

When Eric Bakeman, the Sunpower computer wizard who created most of Sunpower's "data acquisition" software, once asked John Crawford in a meeting about the way a "real company" did something or other, everyone in the room knew exactly what he meant: William himself chuckled, even though the small-engine program would be a likely candidate for summary execution if anyone from a "real" company managed to get control of Sunpower, even if only for an afternoon: when a company operates on a shoestring the way Sunpower does, the first thing that men in suits do away with is anything that is costing money rather than bringing it in.

Sunpower's *own* board of directors has on several occasions issued edicts declaring the small *machina non grata* until it attracts some funding, edicts that held until there was a temporary slowdown on some other project, and rather than paying people to do nothing (as William would rather coyly describe it), why not have Chuck and Al Schubert do a little work on the small engine?

To William, this is his almost sacred role: keeping the bean counters (whether they work for the U.S. Department of Energy *or* for Sunpower) out of the technical side, and letting the engineers do their work free from outside interference. "Stick to the knitting," sayeth William, not countenancing for a moment that bean counting *is* a bean counter's knitting.

If banishing the bean counters to some hill of beans doesn't work, then William T. Beale, president, technical director, and chairman of the board of directors of Sunpower, Inc., a man who reserves his iciest disdain for anything even remotely "corporate," proffers a business-schoolish corporate answer when the question of money for the small engine comes up. "It makes *perfect* sense to fund the small engine internally," he says evenly, "because it permits us to be in control of our own destiny, rather than being controlled by the destiny of the sponsor. Sunpower doesn't need to spend *less* on internal projects; it needs to spend *more*. Look what happens when we tie our wagon to one of these corporate *behemoths* that then promptly gets into trouble and the trouble trickles down to us. Jobs like the small engine should be Sunpower's 'Declaration of Independence.' "

It is a testament to William's persuasiveness as a constructor of sensible and wise corporate policy (even though he despises most corporations) that this sort of argument has carried the day more often than not; William, when he is in the mood to do so, can talk almost anyone into almost anything. When Sunpower was courting a Scandinavian manufacturer for a refrigerator project, a quip around the offices referred to the difficult task of "selling refrigerators in the Arctic Circle." "Gee," said someone in a flash of insight, "that would be William."

If promulgating a wise and sensible plan with the company's best interests at heart (that happens to include support for the small engine) doesn't work, William might try "volunteering" his time. "Just as a matter of interest, I'm not billing any of my time on the small engine right now," he said once in a Monday meeting, "and I promise not to use a single *drop* of company resources, scrounging *only* the most useless and abandoned bits of string and baling wire, so as not to adversely affect the company's momentarily precarious bottom line."

To which Hans Zwahlen added, helpfully, "And you won't use any electricity, either, right?," prompting William to agree to "work in the dark, if need be, until I can work by the light of the *copious* amounts of electricity produced by the small engine itself."

If this doesn't work, William will simply pull rank. It is, after all, his company.

His company, his corporation. His idea. And while the company perhaps was, and still occasionally is, small, the idea has always been large, global, even universal. On good days, when things are humming along, and the beating of the bushes has gotten some money men (always in suits) to troop through and see the working hardware, on days when there is plenty of working hardware around—on good days, the company can look as though it is up to the task, can look up to its goal of changing the world as we know it. On less good days, when William is emphatically supposed to be chained to his desk and beating bushes and the test cells are painfully silent, the whole enterprise can look ridiculously small.

To everyone, that is, but William Beale. Even in the very beginning, William had no patience for thinking small, although it must be stressed that William seldom thinks big in any conventional business sense. It would be stretching things to say that William ever thinks in any conventional business sense for any length of time. No, the "big" that William has always envisioned is a *productive* big, Sunpower as a huge beehive, a world of industriousness and productivity based on the gospel of thrift and decency, a world where the obviously solvable problems are solved quickly and efficiently, so that valuable time and resources may be devoted to the much harder tasks. Changing over from a fossil-fuel-based energy system to a solar-based one may be a challenge in *logistics,* but it is the thrifty and sensible and decent thing to do, and, of course, the technology is there—the technological challenges are, for all practical purposes, nonexistent. The only problems are failures of imagination. At least according to William.

This makes him sound like a fantasist, given how large the tasks at hand loom. Yet no one would ever refer to him as a dreamer in the conventional sense of having one's head in the clouds, woolgathering, thinking thoughts unbound by reality. (Instead, he can seem shockingly concrete, the sort of man, as was once said about him, "who, instead of reading a magazine, will calculate the number of watts consumed by the lights in a waiting room.")

William is "imaginative" only in a special, narrow sense, which often misleads people, or at least confuses them, since the word "imagination" can also be applied to a particular facility for *envisioning* something—something concrete and tangible. It is particularly telling that William's insight that ultimately led to Sunpower—on that day in front of a mechanical-engineering class when he thought of a conventional Stirling engine with the cranks sawn off—occurred to him as a modification of hardware, a concrete change of a tangible object that suddenly had lots of tangible possibilities. Because William, although it sounds oxymoronic to describe it as such, is singular not for his imagination but for his ability to *envision reality*. He is rooted, earthed, in hardware, an attitude the poet William Carlos Williams once described as "no ideas but in things." Dr. Williams helped change the face of American poetry; Mr. Beale wants to change where we get the electricity to read it by.

To name this quality in William—a vision for a particular change in the world we live in—is one way of trying to make him explicable, something that many people at Sunpower have tried to do. They have tried to explain, to understand, him as a way of accounting for the two most singular aspects of Sunpower's quarter century of existence: First, how has it managed to exist, held together by broken string and bits of baling wire (the two all-purpose Sunpower names for materials), for so long, without the benefit of any conventional measure of success? And second, how has it managed to exist for so long and still not have achieved the goal that is the company's very reason for existence, a real-world marketable machine?

That the short answer to both is "because of that rumpled man over there tinkering" is unsatisfying, but it *is* the answer. Sunpower exists because of William Beale, and whatever the failures and missed opportunities and unhappy decisions that have resulted from that fact, it is true. No one else would have done it. If Sunpower ever wins big, ever hits the home run it has been trying to hit for a generation, William Beale will deserve the portion of credit reserved for those who make something happen, who see a world and then set out to make it. He is not the best engineer in the place, and he is not by any definition the best businessman. But if Sunpower changes the world, it will be because William Beale made it possible, but it will happen only if it happens on *his* terms.

Changing the world is a role that has occasionally been reserved for savants, or for visionaries, but perhaps most of all for those who have a thirst for winning the game as it is played, not a thirst for changing the rules so that the field is level. John D. Rockefeller, to mention someone as unlike William T. Beale as oil is from sunlight, saw a societal need and filled it. That Rockefeller became the richest man of his time was not in any way a pleasant but unintended consequence; it was in fact largely the reason for what he did in the first place. Oil happened to be the substance with which he did it. For Henry Ford, it was assembly line production of automobiles; for Steven Jobs and Stephen Wozniak, a computer (the Apple) for the common folk (but available only from them). For Sunpower, the idea has always been something for *everyone,* to change the way the world *is,* for the benefit of *all.*

From *Sunpower Newsletter,* number 1, typed on New Year's Day 1974:

What SUNPOWER is about: I remember once hearing an executive of Curtiss-Wright say to a Junior Engineer, "Remember, young man, this outfit is in business to make *money,* not airplanes." When I heard that, I took an instant dislike to the executive as well as Curtiss-Wright. Making money is, in my opinion, just another way of doing a service that is worth something to others. If SUNPOWER does something that is of value to others, it will probably make money. If it doesn't do a service, it *shouldn't* make money.

Anyone who has ever spent any time at all at Sunpower would recognize in these words the vintage William Beale—the reference to a "Junior Engineer," which dates William's bureaucratic vocabulary as being of the 1950s, when he was in engineering school (Caltech '53) and big corporate R & D operations were becoming the norm; the incredibly apt quote from the Curtiss-Wright executive that is against *everything* William Beale believes in; the "dislike" for a corporation (he dislikes most corporations, mostly for *being* corporations, synonymous to him with "waste" and "sloth," but also because of their tawdry obsession with money); and most particularly, the act of defining Sunpower as being "about" something in the first place: no one ever asks what IBM is *about.* But the most Beale-like touch in the newsletter is the word "shouldn't," underlined by William Beale on his wife's old manual typewriter on New Year's Day, 1974. It is as clearly a "shall not," as if it were carved in a stone tablet. If Sunpower doesn't do something of service to humankind, it shall not make money. It would be *wrong.*

This is about as pure as one can get in any realm; in the realm of capitalism, it is the sort of statement that simply doesn't compute. Men in suits don't think this way, and they are made nervous, suspicious, unsettled, by someone who does; they don't like to be *near* such thoughts. In the world of product research and development, the world where "cutthroat" is a compliment and using your best friend's guts for garters considered not only fair play but smart business to boot, someone out to do good is assumed to be up to no good, because the only way to do good is to win at the game.

And here comes William, the sort of man who is saddened at seeing a bumper sticker saying, "Whoever dies with the most toys wins." "What sort of attitude is *that?*" he asked once in a technical meeting with his trademark half chuckle and quarter shake of the head, to which someone murmured, "The attitude of our *competitors.*" But William isn't like this, and when he tries to be, he does it badly, because he tries only halfway. His heart is simply not in it.

His heart is in a different place altogether, a hybrid world that is not quite business, not quite engineering: a world where what is good for one is good for the other, business making possible research that improves everyone's lot. The world William wants to live in—wants to *make*—is a world where the interests of capitalism and community coincide: where what is good for humankind is what is done. When Sunpower enters a period of tight or nonexistent money, he tries with a vengeance, with all his heart, to become cold and calculating and businesslike, at the very same time he is grievously stricken by what being cold and calculating means, and how it isn't right. He listens to the money men (even though at Sunpower itself, the "money man" is in fact a woman, Accounting Supervisor Debbie Barron), to the board of directors, to Projects Manager John Crawford, and makes a "business" decision and resolves ever after to play the capitalist game like John D. himself, but he simply doesn't have it within him to do so. For relief, he will sit down at his work table and sketch out improvements in some machine or another. Or he will go and work on the small engine, exactly the sort of thing that he isn't "supposed" to do. It is a way for him to work for the Sunpower where the personal is not political but technical, and where no decision is made for financial reasons alone.

Sunpower as a company has thus evolved in ways that visitors from the outside world often find incredible, even—or especially—the companies with which Sunpower does business. To take one example, while the engineering staff is divided, as engineering staffs the world over are divided, into "teams," the analogy that Sunpower in action conjures up is not the world of, say, professional sports but instead that of something more like a high school gym class, where a lot of picking and choosing goes on. This is particularly evident at the Monday morning meetings—the general meeting at ten o'clock, followed immediately by the technical meeting, populated (mostly) by the engineers. The idea behind the Monday meetings is peculiarly Sunpower-like: everyone in the company gets to know what is going on—what projects are up, which are down, who did something marvelous, who had a setback, what that very loud noise was late last week, where the week's visitors will be from. But most corporations—particularly corporations in the 1980s that admired and sought to emulate the lean, mean, no-

nonsense, no sentimentality, bottom-line aggressiveness that marked those years—like to imagine themselves as too busy to congregate, to commune. Sunpower strives, in fact, for the opposite, a sometimes self-conscious reminder, each week, of the company's humble roots and noble aspirations, a place of equality and teamwork, where good sportsmanship guarantees winning, and no one wishes (at least overtly) for it to be otherwise.

Elaine Mather, whose title at Sunpower is administrative manager, but whose roles encompass virtually every aspect of management that doesn't involve actual engineering, runs the Monday meeting from a vantage point in the corner of the large, dark, open space outside of William's office—a wide spot in the hall gloomily lit by a low-wattage, high-efficiency, light bulb in a shade clamped to the drop ceiling. The floor that everyone stands on was sanded and refinished by William's children when Sunpower undertook to spruce up the old, run-down building that would be its new headquarters. It is hard not to imagine William's son John raising the grain in the scarred oak floor, helping make a company happen as the present-day company straggles in. There have been times at Sunpower when the crowd at the Monday morning "beating" overflows down the hall; there have been times at Sunpower when the meeting could be held in Elaine's minivan. But the meetings are held each week, a gathering that begins with all at the company nodding to one another, saying good morning, inquiring about the weekend, asking what's new. Attire is casual; in the summer, anyone not wearing sandals looks dressed up.

A few minutes after ten, Elaine says, "Ohhhhkayyyyyy," the Sunpower equivalent of gaveling to order.

William props himself against the wall outside his office door for the meetings, hands in his pockets, eyes off into space; Peggy Shank, his amanuensis and assistant, sits in a chair and takes the minutes. Elaine surveys the gathering and each week looks as though she didn't quite know where to begin, managing very well to convey the sense that this is a sort of . . . impromptu affair, nothing formal, as though it would be a splendid idea, with everyone gathered in the hall *anyway,* to catch up on some company news. "Let's see," she will say, "let's have project reports." Even though all but a rare few meetings begin with progress reports from the various teams, Elaine perfectly captures the sense that this is *not* a meeting but a discussion, a gathering of equals, not a summoned-to-order assembly. She looks around, saying, "Who wants to go . . . ?" Her eyes alight on a project manager, and her eyebrows lift. "Abhijit?" And so another week at Sunpower begins.

The reports at the Monday meeting fall into three main categories: engineering projects; current and prospective business matters; and anything else. The engineering projects are further divided into broad categories of coolers, engines, and other projects; the business matters range from who has neglected to turn in a time card to what large and rich corporations are mildly, actively, or not at all interested in what Sunpower has to offer. The anything-else part of the meeting is where Sunpower really shows its roots. Parking-lot courtesy; the soft-drink machine; extra vegetables from someone's garden; the news of the world (one week's minutes led off with "Gorbachev ousted in Russia," followed by "Don't use the lower parking lot anymore"). The bulletin board "discussion," which sounded for a moment as though it would pass out of the company's consciousness without comment but ended up as a topic of conversation and negotiation over a period of several weeks, is a perfect metaphor for what the Monday morning meetings can be.

It started with a cartoon that someone put up on the main bulletin board at Sunpower, the board right outside the conference room, the board that virtually everyone passes a dozen times a day and that is exactly the sort of place to pin up something clever, something funny, something newsy, something noteworthy. It was also the likely place for cartoons. And given the sort of place Sunpower is—egalitarian, peacenik, working on a new world order long before it ever occurred to George Bush, the cartoons tend toward the frankly left-wing. And then someone put up a cartoon that made fun of peacenik egalitarians in the context of the slaughter of Chinese students at Tiananmen Square, to the effect that a few guns used judiciously in the cause of freedom might have made some difference. And someone else took it down.

In microcosm, it was exactly the sort of trip-up that happens all the time to liberals; in the dawning age of political correctness at a place where political correctness might well have been born, the removal of a cartoon that expressed politically incorrect views raised disturbing issues. After all, the bulletin board often featured cartoons that poked fun at gun owners, Dan Quayle, conservatives, big business. So when the shoe was for a moment on the other foot, someone took the cartoon down and precipitated the bulletin board crisis at Sunpower.

"Now, we have a *slight* problem with the bulletin board," begins Elaine, and as she describes the scope of the problem, her voice takes on its most patient and forbearing tone, a tone appropriate to exhibiting sincere sensitivity to all, even—*especially*—to gun enthusiasts. "Now, we have to remember that everyone is entitled to her own opinions, and freedom of speech requires that we respect *everyone's* opinion, everyone's feelings. . . ."

The mention of "freedom of speech" catches William's ear, and he pushes himself off his door frame to speak. "Yes, you know, it seems to me that the issue here isn't guns—I've got a gun myself—" (this confession raises some eyebrows among those who have always assumed that someone as bent on world harmony as William Beale wouldn't dream of owning a gun)—"the issue here is whether anyone feels the right to simply censor what someone else wants to say; so the point isn't guns or no guns, the point is how we can treat each other with respect." William leans back against the door frame and cleans his eyeglasses, the matter, he thinks, closed.

Not so fast. As William perks up at the mention of any infringement of freedom, so others at Sunpower perk up at the chance to debate an issue, a frame of mind that can make meetings take hours. One engineer points out that there are all sorts of limitations on free speech—what about speech that offends someone? Getting into the spirit of things, another says, "Yes, what if I wanted to put up a photo from a skin magazine?" This thrust is immediately parried by someone who points out that there is nothing offensive about a woman's body, which leads into the issue of exploitation, and in a matter of minutes there occurs an interesting and quite complicated discussion about the nature of offensiveness—doesn't being offended suggest that it's one's own problem? "Well, we're *not* going to be putting up pictures from men's magazines," says Elaine, drawing a line in the sand. "What are we going to do about this, though?"

Suggestions: no one can post anything on a bulletin board except official business (shouted down); everything posted must get administrative "approval" (shouted down in a blaze of niggling over who would be the bulletin board czar); an area on the bulletin board for posting "offensive" materials (denounced as unworkable, with several people pointing out that what is offensive to one is pleasing to another—the skin magazine issue. "No skin magazines," says Elaine); no rules or limits whatsoever on the bulletin boards—free speech in its purest form. ("Oh, so I could put up a swastika?" "Wouldn't bother me." "No swastikas," says Elaine.)

The issue is left, for one Monday morning at least, unresolved, other than an exhortation from Elaine to be sensitive to others. What finally happens, over the course of discussions at subsequent Monday meetings, is nothing short of amazing—to the rest of the world. At

Sunpower, it is classic. There will be two bulletin boards, one outside the conference room and one in the hall leading to the kitchen. Anything posted on the conference room bulletin board that someone deems offensive can be moved—not *removed*, but moved—to the kitchen hall bulletin board. A shortcut would be posting anything remotely offensive on the kitchen hall board to begin with, a board that will permit free speech at its purest. And everything posted must bear the initials of the posting party—if you have a bone to pick about something posted, find the person and pick it: no anonymous postings allowed. Lyn Bowman takes one last stab at it. "Can we move something that *isn't* offensive to us from the kitchen board to the front bulletin board?" This query earns one of Elaine's looks. William thinks this is a grand compromise, and says so: free speech *and* respect for others. A winning combination, one that captures perfectly—oh, so perfectly—what Sunpower is all about. As a bonus, the final "consensus" didn't really take much time at all, at least not when one considers the issues involved as free speech and respect and working and living together. Put that way, Sunpower is very efficient indeed. Put another way—an hour or so, total, of the entire company's time spent discussing whether removing a cartoon pinned to a bulletin board is an infringement of human rights—it's an open question.

Because it is not just the bulletin boards and the soft-drink machine that get such perusal, such scrutiny, at Sunpower. The whole gestalt of the company is such that to survive—at least in William's view—it must not only take care of the knitting but do so in a way that is humane and humanitarian and decent, yet also frugal and hardheaded and businesslike. When, for business reasons, or as a result of fate, or bad luck, or bad decisions, this cannot be accomplished, it grieves William deeply, both in the particular—having to, for example, lay someone off or cut someone's pay—and in the general, in that layoffs and pay cuts are inhumane.

So there generally comes a series of exhortations, often in the form of memos, outlining where Sunpower is going and reminding everyone of just what Sunpower is about. William's memos are printed out on the dot matrix printer, rather than the laser printer, because the dot matrix printer has the appearance of thrift. When money is really tight, one of the memos will inevitably be about remembering to turn off lights and not fiddling with the thermostat.

The memos from William first make the case for a quick shift into being more "businesslike," to playing the game with the outside world by the hard and cold rules by which dogs eat dogs; the memos then appeal to everyone to put the greater good of the company and its mission ahead of individual concerns. It can seem contradictory—dog eat dog, but in a decent and humane way.

This is the sort of discussion that often takes over the technical meeting, the gathering that follows the Monday morning meeting, and which is intended to focus purely on technical issues. But because the tech meeting brings together the employees at Sunpower who are most in tune with what Sunpower is about (whether they admit it or not), the discussion can quickly move rather far afield. It is perhaps the most interesting place *culturally* at Sunpower, if "culture" is defined in a very narrow sense.

The idea behind the tech meeting is simple: gather all the engineers together to review technical problems, and brainstorm solutions; it's also a time to reallocate personnel, discuss scheduling and assignments for the drafting room and the machine shop. On several occasions over the years, a schedule has been drawn up for engineers to make presentations on their particular areas of expertise; on fewer occasions, such presentations have actually been made.

When the talk is of technology, William shines; he would rather talk about energy conversion than almost anything else, since he loves to take chalk in hand and sketch out some methodology or strategy that was once used to solve some problem, or that will solve some current problem, or that is simply interesting. "I was thinking yesterday . . ." is typically the first thing William says in a Monday tech meeting—new engineers at Sunpower have sometimes considered actually driving out to William's house on a Sunday to see what he is *doing,* not believing that he could be doing second-order analyses on a *Sunday* at home. (Often he is not; he is doing second-order analyses on a Sunday at Sunpower.) William's life and world can seem—especially in the tech meeting—simply the Stirling cycle.

As new engineers become less new, they often reconsider their initial impressions and begin thinking that the *only* thing William does is think about work, but this is not true either. He reads ("I was think-ing yesterday about something I read in the biography of Rudolf Die-sel. . . ."), he fishes for bluegill ("I was out at the pond yesterday, and it occurred to me that the figures for biomass conversion that everyone uses . . ."), he cuts brush ("Wanted": reads a note on the bulletin board one morning, "someone strong to clear brush; hot work, low pay, no benefits, miserable working conditions." "Sounds like *Sunpower,*" noted one wag).

He lives a life, but it is the life of the technical mind. Video games baffle him; once, seeing Jarlath at the keyboard of the sleek new computer bought to perform finite-element analysis, William came over, animatedly asking about the new heater head design. Jarlath was actually at that moment playing a computer game in which gorillas atop city buildings bombard each other with bananas. William watched for a few seconds, all the enthusiasm gone out of him, and then he laughed his little half laugh and lost all interest. He would never chastise anyone for something like this; he just doesn't understand it. He dislikes dressing up (foolish ornament), drinks coffee so rarely that one cup makes him perky, has no habit of small talk. With the rare exceptions of time spent with the Unitarian fellowship or the several environmental organizations he works with, William Beale lives in the environment he has created, and lives it personally and viscerally—but lives it through a vision of the possibilities of technology.

On the day in 1991 when Sunpower, as a result of a confluence of two of its more complicated funding arrangements, shrank into an enterprise with half as many employees as it had had the previous day, William actually missed work—out, he reported gruffly two days later, sick. It left Elaine Mather to be the bearer of bad news. Some at Sunpower, watching friends (and in two cases, *spouses*) pack up and head for the unemployment line, thought this was a form of avoidance; Elaine thought that William was sick all right, sick over what was being done that day. Those who survived the layoffs took pay cuts; some took pay cuts and time cuts as well. It was a business decision plain and simple, Micawber's rule for happiness applying at Sunpower as it applies everywhere else, and Sunpower, on that day, was out of money in a way that Micawber couldn't even dream about. Businesses lay off employees when money is tight all the time; they "downsize," to use a term particularly revolting to William. On paper, the layoffs were the right—in addition to being the only—thing to do, but they weren't the *decent* thing to do. In William's world, money flows to those who need it and deserve it, those who are doing, trying to do, the right thing. He wants two things equally badly: good work and the reward of virtue, in a world where the big boys want only good work. John D. Rockefeller would have consumed Sunpower for a midnight snack and used the skeletons for toothpicks.

In capsule, this tenacious attempt to hold on to what Sunpower is "about" is also, then, the business history of Sunpower, a history that sits as an overlay on the engineering history of Sunpower, although it is seldom possible to say in any given example which was a by-product of the other. The two histories, taken together, are a professional history of William Beale.

There is, for example, Sunpower's long (and still-present, although in mighty small form) relationship with the Third World, in the form of Stirling water pumps and Stirling solar generators and rice-husk-powered Stirling rice huskers and water pumps again and solid-fuel Stirling generators that run on cow chips and the latest incarnation, yet another water pump. To William, the Third World was one of the "obvious" places for Sunpower to succeed, a textbook example of Sunpower's providing something of value (water pumps, rice huskers) to those whose lives would be enriched by it. Win-win, right off the bat. If one factors in the global benefits of replacing environmentally harsh practices and fossil-fuel consumption, even those people who have nothing to do with the enterprise win as well, so it's win, win, win. Solar-powered generators in the Third World would limit deforestation and desertification, would limit global warming, benefiting everyone yet again. Dictators wouldn't need to dictate, because their citizens would be happier, healthier, better fed.

There was in 1974, and there is today, a desperate need for economical and environmentally benign power and water sources all over the world, simple devices that pay huge dividends in the alleviation of human suffering. Furthermore, because the Third World hadn't already invested in a fossil-fuel-dependent, capital-intensive, corporately entrenched infrastructure, the transition would be not only effective but efficient—what the First and Second worlds would have done had they had the advantage of a few hundred years more of R & D.

It all sounds so sensible that it can be baffling why it hasn't occurred already. If Sunpower had a warehouse full of rice-husk-burning rice huskers available at a reasonable—read, "Third World"—price, it could perhaps, or even probably, sell them, and the Third World would beat a path to the Byard Street door. Instead, what Sunpower had was a design for such a machine, and the expertise to make a production prototype. Many companies and agencies would be delighted to sell the finished machines, or try to. There is, though, no one to make them, no one with the capital and the vision and the sense of community to front several million dollars to manufacture a warehouse full of such devices. (In this case, it is clear enough, though, that if no one on earth wanted a rice-husk-burning rice husker, it wasn't because of William: a group of former Sunpower engineers who spun off into their own company in the mid-1980s took this design with them, and they have had no more luck than Sunpower.)

Sunpower gets calls regularly from the World Bank, from AID, from the UN, from foundations large and small, which all want to help fund distribution of what sounds like a miraculous device: a simple machine that makes life in developing countries easier, cleaner, and more productive. None of them are engine manufacturers; they want something in a box. William, working the phones one afternoon, receives a call from the president of an international foundation who has heard about the small engine: how much does it cost?

"Right now," says William, "because it's still in the prototype stage, ten thousand dollars."

His caller laughs, and tells William that he is off by "two orders of magnitude." William has heard this before; everyone wants a hundred-dollar small engine, but few want the $10,000 version that must exist before a cheaper model can, and *no one* wants the first engine—the $100,000 one—at all. Abhijit Karandikar, who over time has inherited some of the duties associated with trying to market the small engine, has a binder in his bookcase full of the records of calls such as this. He could sell a shipload of small engines tomorrow, but he can't find the hundred thousand dollars or so necessary to fund the development that would make manufacturing a shipload possible.

John Grandinetti, who owns and operates Grand Solar, Inc., in Hawaii, and who sponsored some of the R & D on the small engine, is a typical small-engine sponsor in many ways, in that he *(a)* had grand plans and *(b)* had relatively little money.

Grandinetti's plans were writ large, even for a Sunpower sponsor; what he bought from Sunpower was nothing less than the license to manufacture and sell a portable, versatile, and rugged Stirling engine in China, where the potential for a remote source of electrical power is huge. The idea of the small engine is an engine for the rest of us, one that would make several hundred watts of electricity, run on a wide range of fuels, and be small and sturdy enough to hold up many hundreds or thousands of miles from the nearest person who might know how to fix it if it happened to break. Grandinetti worked on setting up contacts at the United Nations, with Chinese governmental agencies, with international development agencies. Sunpower kept him posted on the progress of the R & D.

What Grandinetti heard Sunpower say over the course of a year of engine development was that things were going fine, that the engine had great potential, that progress was, in William's phrase, "too good to be true." The design suggested a small engine that would produce something on the order of 300 watts; Sunpower's own publicity materials on the small engine stated "200 + watts"; John Grandinetti understood this to mean that when he flew from Hawaii to Ohio to pick up the engine and take it to the United Nations for a demonstration, he was getting an engine that produced somewhere between 200 and 300 watts.

What he in fact arrived to see was a very nice small engine, a fairly compact, fairly robust package emblazoned with the Sunpower logo, that produced 180 watts. The engine had generated nearly 250 watts on occasions during testing, but when finally configured with its electronic controls and its solid-fuel burner, it did not.

Grandinetti had in mind a picture based in part on how he interpreted what he had been told. That Sunpower understood *exactly* why the engine produced 180, rather than 250, watts was of little consolation to him (it had to do with time and money): he wanted the rest of his watts, and he wanted them *now,* before he left for the United Nations. The several hours of rather tense conversations that everyone at Sunpower knew William and Grandinetti were having in William's office that day might as well have been broadcast over the intercom system: everyone at Sunpower knew what was being said, as well as they knew what Grandinetti had in his pocket—a check representing the last license payment due Sunpower, in exchange for borrowing the small-engine prototype. The engineers and technicians in the lab (some hundred yards and several sets of doors away from the office) stood around trying to divine the exact moment when Grandinetti decided to hand over—or not—the money. "Well?" one engineer, passing through, asked the group. "Not yet" was the reply. It did happen a little later.

If William Beale is guilty of anything in this world, it is of always wanting things to be better, a hard point to argue against, although John Grandinetti did so. What Grandinetti also does is perform a sort of risk taking that most of the time only William finds acceptable—he pushes at the envelope wrapped around him by his own company.

No small-engine sponsor, or licensee, has ever presented Sunpower with the *real* money necessary to make it into a manufacturable machine. Instead, sponsors have put up part of the money, and William has scrounged the rest from other Sunpower projects. William sees sponsors—Sunpower's only customers—as funding the development of a technology, not a product, and there is always the expectation on William's part that sponsors will be rewarded at some point. Sunpower's customers, however, generally have different fish to fry. They tend to want what was promised and nothing else. If Sunpower could charge a royalty on being told by tetchy sponsors, "Better is the enemy of good enough," Sunpower would need no other income. But the fine line between good enough and not good enough is usually buried deep in the dynamics of a machine that responds in big ways to small changes.

In the case of a project like the small engine, the fact that it has been starved for money for so long means that it can never be technologically parallel to better-funded projects; because it isn't technologically parallel to better-funded projects, it has a hard time attracting funding, and it scrapes along on string and baling wire, with William volunteering his time and Chuck and Al helping out when there is nothing else billable to do. In microcosm, this is pretty much the way that Sunpower ran before it got its hands on some real money: in one sense, the small engine forms a sort of "real" Sunpower, Inc., hidden inside a corporate Sunpower, Inc. And it is one of the reasons that Sunpower runs into trouble with its bigger money sponsors, because they tend to think of contractors as providing a product, not a process. Sunpower, William has said over and over, isn't selling products, though; it is selling its expertise. This is where the Sunpower vision, with what Sunpower is "about," comes up squarely against the wry "Golden Rule." ("Whoever has the gold makes the rules.")

The other problems are equally perplexing, and unforeseeable in a theoretical sense: they become clear only when someone like William Beale tries to do something. He doesn't want politics and gabbing and gassing. He wants *hardware*. He has no patience (although he should have learned it by now) for the mixed message that the developing world sends off and receives. "If this technology is so ideal," William was once asked by a bureaucrat from Asia, "why don't *Americans* use it?" Many developing countries are wary of being proving grounds for technology that isn't accepted in the developed countries at the same time they are being enticed, often by the governments of the developed countries, to make themselves in the American image: with cars, highways, power plants, hydroelectric dams, factories.

There is deep suspicion in developing countries of "ideal" and "appropriate" technologies that don't exist anywhere else; they sometimes see this as a sort of chauvinism—good enough for *you,* but not for us—but more often, they see it as a limiting, rather than an expansive, "improvement," because it would slow down the *real* development, the one with the cars and the refrigerators and the television sets, and the consequent oil refineries and power plants necessary to run them. This response was at first genuinely baffling, then positively frustrating, to William Beale: "if you make something that is of value to others, . . ." doesn't take into account the possibility that those others *still* might not want it. The whole scheme, from simple and cheap water pumps to rice-husk-burning rice huskers, makes perfect sense, but it makes perfect *engineering* sense, this in a world where the last engineer with any political power was probably General Leslie Groves, the man who built the Pentagon (ahead of schedule and under budget) and oversaw the Manhattan Project. Before him, it was Herbert Hoover (Stanford '95).

Many a company has been set up, and has then foundered, on the notion of opening up the Third World with products of every stripe. ("China! One *billion* consumers!") Back in the Second World, successful start-up companies are generally those associated with "must have" technologies or products—videocassette recorders and huge, flashy basketball shoes come immediately to mind—or those that carve out niche businesses in the interstices between huge corporations. Some also do it with new—brand-new—technological developments that stake out uncharted territory, such as biotechnology.

None of these categories contain a company that wants to *replace* one technology with another, especially a technology where the allegedly obsolescent one is as huge and ingrained as power generation. Electricity generation isn't "must have," because everyone already has it; it isn't a niche—in fact, it's quite the opposite. And it isn't uncharted territory: there is an electrical outlet in virtually every room. So if the Sunpower scheme is ever to work, it needs some other edge. So far, that edge has been William Beale.

This is perfectly true even though the company has not been a success by any conventional measure. It is true because William Beale, through sheer force of personality, has kept the company warm through each cycle of pessimism and bust, through each downsizing, and through each technical and financial dead end. If Sunpower had folded up its tent a decade ago, much of the commercial potential of the Stirling engine would have been folded up with it. William alone hasn't kept interest in the Stirling engine alive all these years, but he has done more than anyone else to keep dragging and tugging it a little closer to commercial viability. It is William who has been there when each of the small opportunities of the past twenty years has emerged that makes the Sunpower dream a little more possible.

There are, counting very broadly, a dozen companies or research groups investigating Stirling technology in the United States (not including academic institutional research); worldwide, perhaps another dozen. Of these, Sunpower is the closest to commercial success. And without Sunpower—without William Beale—there would be no one nearly as close, and no one thinking along the lines that are making Sunpower close, because so many of the engineers working on Stirling technology have, at one time or another, passed through William Beale's orbit.

It would be grand to say that this was the result of an overarching vision, but it is probably the result of circumstance: it is impossible to say where, if Sunpower had been given a blank check in 1974, it would be right now—what, if anything, the company might be making. But it is safe to say that the company would not be doing what it is doing today, merely on a larger scale and in more comfort, because the original vision was much different. The original vision was simply what William Beale looked around and could see.

Things like the box in the basement—perhaps the first conceptual product of Sunpower, and one that still exists as the small-engine project; rather than utility-scale engines, William envisioned a home-scale engine, starting the company as he did in the decade of self-reliance and woodstoves and the Alaskan Pipeline. If everyone used one-third less energy via a Stirling engine in the basement, there would be no need for a Trans-Alaska pipeline at all. "It takes no imagination to convert this [idea] into potential profit for Sunpower!" is another telling line from the first newsletter, one that follows an arithmetic demonstration of why a three-kilowatt Stirling generator in every home made perfect sense. The next sentence is also telling: William acknowledges that they don't know how to *make* a three-kilowatt generator. Yet. So they set out to try.

William as a tinkerer nonpareil will always be the foundation of Sunpower legend. Make something, try it, take it apart, and improve it. Make a change (or six), try it, take it apart, and fix it. A classic William Beale design drawing is one that contains, like an archaeological survey, a half dozen different possible ideas superimposed, one on the next, on one sheet of paper. It is an undisciplined way of working, but a fertile one: some people cannot look at any object without thinking of ways of improving it. For someone so single-minded, William is in a strange way *open-minded,* willing to try all sorts of things that occur to him ("Cut the cranks *off*") and seeing what happens. Once, on the small engine, the power piston came out of the machine shop in a shape not specified: rather than being straight sided, like a soda can, it had the gentlest taper, one round end slightly smaller than the other. To most people, this would be a mistake. To William, this was a design improvement that would help the piston run in its proper place in the cylinder. For several weeks, William considered this to be a simple solution to a common and vexing problem with Stirling engines—how to "center" the piston in its cylinder, and he went around suggesting tapered pistons. When, for various reasons, it turned out not to be the solution he thought it would be (it was in fact harder to make a piston with the proper taper than it was to make one with straight sides), he simply went on to something else. Some ideas stick, some don't, but if you have an inexhaustible supply, this is not a problem but an advantage.

William has long run, or tried to run, the company in the same way; think about it, visualize it, listen to ideas and suggestions, and do something. If it doesn't work, try something else. In hardware, this is known as experimentation; in business, this is known as trouble. For example, the most precarious position at Sunpower from its very inception has been that of top businessperson, the person charged with the responsibility of running Sunpower like a business. The list of people that have held this job, formally and informally, is long indeed, and the current officeholder, Projects Manager John Crawford, is notable not least for having held the spot longer than anyone. That is, a little more than a year, and counting. (When Crawford took the job on, one engineer said in genuine puzzlement, "Really? I didn't know Crawford wanted *out*," the assumption being that the business job was the fast track to storming out of Sunpower in a huff, never to return.)

The position is so precarious because its occupant must do two things simultaneously: be businesslike while remaining true to the Sunpower ideals, all the while getting advice and strategy from William, who has run the company himself, with varying definitions of what this means, on many occasions, usually just long enough to remember why he detests it so. In a practical sense, the businessperson's job means getting money for Sunpower from the hardest and most unlikely sources, while eschewing the easiest and most likely sources. Sunpower's funding comes from loans, sponsored research, and license agreements; it doesn't come from offering stock, making joint ventures (unless Sunpower controls the venture), or selling a portion of the company to someone who wants to invest. "Well, you know what so-called 'venture capitalists' want to do," William says, acidly; "they want to own the majority of the company and bring in their own people to run it as they see fit."

That someone who owns something would want to run it shouldn't be anathema to William—it is, after all, what *he* does—but he knows that money men care only about money, not Stirling engines ("this outfit is in business to make *money,* not airplanes"), and to William this is a recipe for failure: one bad patch, one bit of bad luck or misjudgment, and Sunpower would be vulnerable to the common business practice of shutting something down because it isn't making any money. If Sunpower were making money, lots of people would want to buy it, but William wouldn't need to sell it, or a portion of it; when Sunpower isn't making money, it is a riskier investment, tempting only to the sort of men in suits who could be counted on to cut their losses and run, and thus would Sunpower end. So, but for a few bits and pieces of stock parceled out to some employees in a scheme in the mid 1980s, William owns Sunpower, and the businessperson on board sells the technology.

Crawford, and everyone else at Sunpower, to greater and lesser degrees, then, sells mostly what are curious forms of licenses to manufacture a certain Stirling machine of Sunpower design, with a fee up front and a royalty on every production unit the licensee makes, if the licensee makes any production units at all. If one had, for example, an agreement such as this that covered, say, a widget on an automobile, and the licensee were Honda, money would roll in for every widget that Honda made. If one has, for example, an agreement such as this for a Stirling-powered widget on a personal computer, and the licensee is Big Computer Company, the money should roll in.

But a license to make something is not a contract to make something, as Sunpower learned to its grief when its top-secret sponsor, a computer concern code-named Cucumber, told Sunpower, in essence, "*Love* the technology, but we've changed our minds about doing anything with it." Cucumber did say, "Thanks, anyway, though." This is unlike Sunpower's other large sponsor, the absolutely unsecretive Cummins Engine Company; all of the brochures touting its Stirling program are noteworthy for not mentioning Sunpower *anywhere*. Sunpower's receptionist, Dolly Walker, affixes a "Sunpower: Stirling Systems Development" sticker to the front of every brochure—*above* Cummins's own name.

Cucumber's change of heart offers a textbook example of what Sunpower has been through so many times as to be emblematic. The small Stirling-cycle cooler (also code-named Cucumber) that Sunpower designed and delivered in units of five and ten to Cucumber, the corporation, were exquisite little machines, performing better than their design specification and at a higher efficiency than Sunpower promised. (A cold computer chip can run faster than a hot one, and in the computer industry, faster is better.) The machines were a tangible embodiment of David Berchowitz's mind: delicate, thoughtful, imaginative. The project was done, by Sunpower standards, within budget and on time—within a few weeks of the deadline and without spending a *lot* of money the company didn't have.

Cucumber got rave reviews at Cucumber, and for weeks there was talk—generally verboten, knock-on-wood sort of talk—about royalties, about a second-generation Cucumber, funded by Cucumber, that would be colder, and smaller, and even more exquisite. There was talk not only of profit-sharing but of profit to share. Money for other projects, money to really work. There are millions of computers in the United States, millions more around the world; if Cucumber, the corporation, went whole hog with Cucumber, the machine, the Sunpower dream of doing something of value to others—even if they were computer owners, not rice huskers—would come true.

So when Cucumber, Inc., said that it thought it would take a pass, Sunpower was outraged, especially when the ramifications of the decision began to hit home. In the short term, there would be no money for Cucumber II, which meant that several engineers and technicians and machinists and draftspersons would have nothing funded to do. Further, since the whole program had been done quietly, secretly, even clandestinely, Sunpower engineers were not even really supposed to tell anyone what they had done. Further still, they had designed and perfected a machine that they then couldn't—this was Cucumber's contention—even sell to anyone else.

That Cucumber's decision was a classic capitalist business decision was even more galling: its people quit working on Cucumber when their largest competitor announced that *it* was stopping research on its own Cucumberish device. If there's no competition, there's no competitive reason to have something, or so the argument went.

Engineers are recruited to Sunpower (when times are good) with the expectation that they will remake the world. The kinds of engineers that Sunpower tends to get are those who are willing to be convinced that this is possible and, more important, those who *want* to do it. At some point in their orientation, though, they look around, around the "morgue" at Sunpower, where the carcasses of dozens of stillborn and long-dead projects lie; they look at the various scars and stains and smudges on the floors and walls of the old building on Byard Street, each with a story of failure, miscalculation, bad luck, behind it; they read, in idle moments, old log books from moribund projects, pore over the engineering drawings of their predecessors' best ideas, puzzle through pages of painstakingly made calculations written by persons long gone from the premises; and most of all, they hear about, or witness themselves, legends and stories of project after project after project that for some reason went awry. GRIZZY. The ten-kilowatt solar. The three-kilowatt air. Fat Man. SPIKE, which blew up in Kawasaki City, Japan. Cucumber, the perfect machine that the sponsor didn't need and Sunpower not only couldn't sell but couldn't even *talk* about. They hear of sponsors that went bankrupt, sponsors bought in corporate takeovers by new owners uninterested in Stirlings, of sponsors who simply disappeared, of sponsors who, in the cold, clear light of capitalism, decided that if no one else had a Cucumber, they didn't need it either.

At some moment, every Sunpower engineer who signs an imposing nondisclosure agreement in Elaine Mather's office and shows up for work—early on the first day, less early as time goes on—every engineer looks around, after a project has been stifled or blown up or trashed, looks around and wonders, at first in the dark watches of the night and then more and more often in brighter and brighter rooms, "Is this *crazy?*"

"I started working on free-piston Stirling engines on my birthday in 1964," William Beale says at the Monday morning meeting on his birthday in 1992, and one can almost hear a dozen engineers doing the subtraction. "That's a *loooong* time ago," says Robi Unger, and everyone else nods vigorously; there are days when twenty-eight *minutes* with a Stirling engine is cause enough to head for the West End Tavern.

The nerve that it took, ten years after William started working on Stirling engines, to abandon his associate professorship at Ohio University and start Sunpower, Inc., seems at once incredibly impressive and funnily, giddily, small: to give up tenure and a regular salary for the uncharted territories of small-business ownership takes gall. On the other hand, William thought that Sunpower was nothing short of a *sure thing*—he said jauntily in 1974 that if it didn't work out in the next few months, he'd be "back teaching Thermo I" at the university, something that he had no intention or expectation of doing. Make something of value to others, and the world will pay you.

The early years at Sunpower are remembered mostly for the technical difficulties, as the time in which William and a succession of other engineers discovered how vexing it was to make something as dynamically quixotic as a Stirling engine behave in a predictable and reproducible manner. In the compressed history of Sunpower, Inc., the company's first decade is often written off as a sort of mad-relative-locked-in-the-attic series of failures, which in fact it was not. Three things occurred in that decade that have ramifications for Sunpower today that are so ubiquitous, so ingrained, that they are taken for granted. In fact, they are the fundament of the company. What Sunpower has that has saved it from extinction is vast tangible experience, the intellectual effort of dozens of men and women over a long period of time, and the tenacity of William Beale.

The initial reliance on trial, error, and trial again in the design, fabrication, and development of engines has given Sunpower an unrivaled knowledge base, earned in *hardware:* the Stirling cycle is seductive in theory but sirenish in practice, and most researchers in the field quickly migrate to either paper or computer studies (much cleaner and easier than hardware) or intensive research on certain *aspects* of hardware design and development—regenerators, heat exchangers, bearing surfaces (for these, a test rig will suffice). Sunpower has had many, many projects, many, many designs and schemes and plans that didn't pan out. As evidence of this, there are parts and parts of parts all over the place. (Chuck Howenstine is always agitating to fix some of them up, hating as he does to see good stuff go to waste.) To some, this might seem negative, but in fact it is one of the company's greatest strengths, as long as it remains in business—if the company goes under, then it will indeed be negative, but while the company survives, it means that there is very little in the field that Sunpower hasn't tried or at least thought about. Its engineers know what works in hardware, and what doesn't, because they've tried it; they know what specifications are within reach when it comes to fabrication, and they know what is a fantasy, because their machinists have told them; they know what sorts of moves will make something run away, or burn up, or burst, or run with decadent inefficiency, because they have run away from runaways, put out fires, dodged flying parts, and divided Power Out by Power In hundreds of times, sometimes on the same machine. When Sunpower engineers go to conferences or conventions, it is rare indeed for them to hear a presentation about which they cannot say, "Gee, William tried that *years* ago." The simple truth is that Sunpower has given up on more designs and configurations than most experts in the field have even thought of. It makes for fewer and fewer false starts and blind alleys as time goes on, because Sunpower engineers start with certain solid information, hard-earned and hard-won on dozens of other projects.

A collateral benefit of this—and something else that sounds negative at first—is the sheer number of engineers who have worked at Sunpower and who have consequently left behind months and years of effort, effort that has been absorbed into institutional memory. It is not just the legends of failure that are shared over the years but also the successes, some of them small, some of them fundamental to the technology, that exist in the repository of ideas that is Sunpower, Inc. This is one of the rarest qualities that a small company can have in any endeavor; in Stirling research, it is crucial, because the approaches that work are seldom self-evident. Schemes for centering pistons, for springing displacers, for designing and fabricating alternators, for making regenerators, schemes that Sunpower engineers are able to adapt fairly quickly and cleanly for their current projects, exist mostly because someone at Sunpower—someone generally long, long gone—figured it out and because someone at Sunpower who isn't gone—generally William or David Berchowitz or David Shade—remembers how it was done. When the company gathers in the driveway or the machine shop each spring for its annual picture, many of the faces are different, but there are also ghosts, ghosts of engineers whose work has survived their time at Sunpower.

But this process of "graduating" so many engineers may have been detrimental to Sunpower in some ways: the same experience and knowledge that is so useful as institutional history would be ever more useful if it were embodied in the engineers themselves, rather than passed down in their absence. In the cases where the departures were involuntary but financially necessary, this is particularly true, and is one of the things that made the 1991 "Sunpower Massacre" so painful for everyone: good engineers with important experience and crucial contributions to make were let go because there was no money with which to pay them. Other departures hurt Sunpower in subtler ways—the 1988 exodus of Bruce Chagnot and Gary Wood, among others, as the result of a series of disputes with William, took away not only some wise engineering talent but some popular engineers as well. It made Sunpower a rather cheerless place to work for a long time thereafter. Cheerless place or no, Sunpower exists as a result of the fact that William has not shown any signs whatsoever of giving up, giving in, or quitting. It is hard to quantify, but his almost unfathomable willingness to press on in the face of what men in suits would call a problematic track record means that Sunpower has continued to exist at times as nothing short of an act of will. Or, as has been said many times in the manner of describing a decree, an "act of William."

William has never been particularly popular at Sunpower, although he is in many ways likable. It just seems to take a bit of distance to appreciate him. There are engineers who find it easy to believe that William goes out of his way to learn their particular vulnerabilities, their particular quirks, and then pokes, probes, at them, as though goading: that William never betrays that anything of the sort is going on can seem confirmation that he is good at it.

Some are driven crazy by William's "sloppiness"—his unwilling-ness to attend to detail, to protocol, to methodology. To the engineers at Sunpower who both love math and are good at it, William's "guess" at a number, or the range in which a number will fall, is all the more infuriating because it tends to be right, or close enough. They may know, rationally, that he is drawing, first, on a body of experience and knowledge that they are years from possessing and, second, on what can only be described as an intuitive sense for what a given stress, a given loss, a given flow rate, will be. But it still can be enough to make someone scream when William will do enough of a calculation—often in his head—for him to say he has "done" the calculation and to suggest a way to proceed. This can make him look like a guesser, a lazybones, or, worst of all, someone pretending to know how to do a calculation when he really doesn't. When he is wrong—and he can be wrong—it is a dismissible event, because someone else has had to do a "proper" calculation to show it, and right or wrong, they now know the answer and they can proceed. When he is right—and he is often right, or close enough—the calculation confirms it, and they can proceed. Few people believe that William pays no attention to events such as this, even if he appears not to. That William is "sloppy" by the standards of, say, the Sandia National Laboratory (an organization so obsessed with detail and documentation that it gives new meaning to the term "anal-reten-tive") is not in dispute; that William is so sloppy as to need to be "rescued" by engineers who do all the calculations is a misunder-standing of the man.

Although he would never call them such, William *does* treat, and think of, many of the engineers at Sunpower as "junior engineers" in an old-fashioned sense, while at the same time he exhorts them to work in, and for, a more democratic, less hierarchical world. Indeed, there are engineers at Sunpower who would simply leave and not return if their titles were changed to anything sounding like "junior" anything. Titles at Sunpower may include (in ascending order) "test engineer," "engineer," "project engineer," and "senior design engineer," and engi-neers of the last three categories may be in charge of the day-to-day conception, management, and implementation of work on million-dollar projects, but William has always reserved for himself—and made no bones about it—the title of "technical director" on *all* projects. He will often defer (in maddeningly backhanded ways) to other engineers on all manner of points, but he will often not. He can pull rank in an instant and in the same breath deny that he has done so, because rank is not what Sunpower is about. Sunpower isn't about *rank;* it's about doing a job. Then he'll tell someone to do it. Once, in the process of exhorting the engineers to be more assertive in a technical meeting, he said, "The one thing we don't need here is yes-men, *right?*" It's a tricky question to answer.

If it is not William's "sloppiness" that gets someone's goat, it might be his seeming arbitrariness; if not arbitrariness, then something else— say, his apparently casual determination that a problem has been solved and that the solution should be implemented straight away. When it appeared that Robi Unger and David Berchowitz had come up with an ultimate solution to one of the most vexing problems in Stirling-machine history, William simply declared that their strategy should be implemented on all the hardware in the place at the earliest possible moment, leaving the details to be hammered out by the individual project managers. He made it sound as easy to do in hardware as it was to sketch out with chalk, the sort of thing *guaranteed* to get the goat of half the engineers in the place, those who never said a thing was done until it was done in *hardware,* not on a chalkboard. But to William the *idea* was the hard part, because while the hardware is always vexing, it is also tangible, and what needs to follow an idea is implementation, implementation by the engineers who are good at implementation— the very engineers who are driven round the bend by William's seeming disregard for the complexities of implementation. It is a subtle point, one that in other fields might almost fall into the category of "honor," or "noblesse oblige," this distinction as to when something can be said to be "done." To some, a job isn't done until there is a calculation, a sketch, a drawing, a test rig, an implementation, and another calcula- tion—it's the way that scientists work in gleaming laboratories when they write papers for learned journals that take two years to accept, or decline, a report. It is the way that Ph.D. dissertations are done: the theory, the math, the design, the test rig. It is the stuff of elegant presentations, replete with slides and transparencies and references to "the literature." It is, in fact, what Sunpower's engineers do when they lay groundwork for patents by discussing from conception through implementation the genesis and the result of an idea in front of a patent attorney's video camera, all the ducks in a row. And here is William, suggesting a trip straight from the chalkboard to the finished product, this afternoon, if possible, tomorrow at the latest. It is not the way they do it at Bell Laboratories.

The problem is, though (and William not only knows this but lives it every day), that there are boxes of meticulous records and half-written dissertations all over the world, carcasses of ideas that ran out of steam, or time, or money before they gathered enough critical mass to support themselves. Sunpower itself is a company that has survived thanks to artificial respiration and last-minute telegrams from the governor more times than anyone who isn't positively masochistic would care to count. Sunpower has existed under the threat of being hanged in the morning for so long that William doesn't have to see the gallows to know it. Who wouldn't be affected by the experience?

So William resorts to what can most charitably be described as "delegation," although such a word implies a hated hierarchy rather than a team or a happy family. The act of delegation implies the existence of something like a junior engineer, while the ideal of Sunpower is equality and democracy. When William's personality is able to rule, to run, a project, it is easy for twenty years of impatience to take hold, and thus for calculations to be skipped, drawings not done, test rigs not conceived, implementations tried, rather than implemented. Projects can get away from the project manager by the sheer force of William Beale's determination to get somewhere, a place he has been trying to get to since before some of the younger engineers at Sunpower were born. Confrontation with William seldom works—David Berchowitz can get away with it in endearing and funny ways (the "fourth" law of thermodynamics at Sunpower holds, "No one can be mad at David Berchowitz"); Robi Unger can make confrontation dissolve (he makes William laugh); Neill Lane can do one thing while simultaneously (or late at night) keeping an option open; Abhijit can diplomatically defuse. Some engineers sulk or try to ignore. Others have been driven to cynicism, frustration, apathy. Some have been driven out the door.

But no one stays, or leaves, without recognizing that William Beale is, finally, a *fact* at Sunpower, a fact of its fate and of its existence. He is a repository at once of the company's hopes and of its history, and those who wish to do something other than repeat it must do so around the central, unalterable fact that William is going to run Sunpower somewhere, going to make something happen in a place and with a technology where everyone—not just men in suits—would have found something more sensible to do. The *Japanese* have very recently all but given up on Stirling engines; their withdrawal from the field was more significant than that of the Swedes or the Dutch or the Nazis or the British or Ford or General Electric or Volvo or the government of Bangladesh. This is the sort of knowledge that Stirling-engine designers have been known to torture themselves with: *I can understand the Swedes, but the* Japanese *have given up?*

But William Beale hasn't; if anything (and this is also sobering to those who think they know him), he has gotten more intense, more focused, more determined, even though the practical effect of this has been to make him seem more diffuse, less attentive to his knitting on occasions when he sees the Holy Grail: the chance for a well-funded, well-executed production machine, with a sponsor willing to pay for it while at the same time nervy enough to give Sunpower the autonomy it needs to make something happen. So Sunpower made some crucial decisions when Cummins Engine came calling, the largest of which was undoubtedly the call to bail out of the NASA project and put its efforts into the Cummins job. Even early on, this decision had its aura of a Sunpower legend in the making: quitting a big and "safe" job (safe, mostly, because it would be years of design and computer work before it reached the hardware stage) for the chance to get right to the hardware. This is very much William Beale, whose patience for anything on paper is epitomized by a candid photograph, taken years ago, of him standing in the machine shop, his hands gesticulating as he shows the machinist what he wants a particular part to look like. No ideas but in *things*.

The Cummins project, though, would start life owing much to the conceptual design for the twenty-five-kilowatt NASA job, for which the National Aeronautics and Space Administration sponsored a design competition. (That NASA is doing such things is part of a long story, but it's the same NASA that operates the space program.) The NASA 25K might have worked (it never made it past the preliminary design stage), but its requirements, and particularly its constraints, were breathtakingly rigorous: "It might have worked," Neill Lane said once, "but I wouldn't have wanted to be in the test cell when they ran it."

So the design evolved; the newly christened "5K team" took what it could use from the NASA idea and made up the rest. And although any design engineer will say that starting with a clean sheet of paper is *harder* than adapting an earlier design, it is also true that if the goal is an ultimate machine, a production machine, every bit of flexibility helps. (The small-engine program, for example, is not helped particularly by using an alternator adapted from an earlier project, if for no other reason than that Sunpower has learned so much about alternator design in the interim.)

Because of this immediate experience, and because of Sunpower's long and tangled history, from the very beginning of the Cummins solar five kilowatt, Neill Lane, as senior design engineer, and Abhijit Karandikar, the project manager for the 5K, were determined to do two things, the two things that they had discerned from their experience and understanding of how Sunpower does—and does not—work. One was to hit their marks—make sure they satisfied the specifications at every step, no matter how impatient that made "certain people," which is how they often came to refer to William when they planned and designed.

The other thing they knew they needed to do was run a tight ship, be methodical, do everything possible to try and make the leap from theory to hardware. When they reported on their progress in meetings, they would take great pains to point out that they were doing things by the book. They would do the calculations first. They would do the computer simulation. They would do the drawings, and all the drawings would be checked by two engineers on the project, and only then would they make the parts. When they made a big change or modification, they would hold a design review, and play devil's advocate.

It is an odd way for a company, and for William Beale (as technical director) to come full circle. For various reasons, most of them financial, the company had spent many years and many dollars on jobs that made the corporation's name a misnomer for machines powered by almost everything but sunlight: jobs for the Gas Research Institute (natural gas), for the Department of Defense (diesel fuel), for General Electric (electricity), for Black and Decker (propane), jobs linked to Sunpower not by the sun but by two more-prosaic qualities—the sponsors had money, and they held out the hope of a production machine.

For in all of Sunpower's history, and for all its technical advances over time, the company had never done one thing: it had never seen a machine leave the place and go into production. So Neill and Abhijit aim for the 5K to be a singular project at Sunpower, one that goes step by step by incremental step from conception to shipment, well documented, well executed, well tested, and, above all, on the money both figuratively and literally. It is a tall order, requiring fealty to both the letter and the spirit of the laws of thermodynamics, to the principles of machine design, and, perhaps most of all, to what Sunpower—that is, William Beale—is "about."

Sometimes they manage to stick to this. And sometimes they don't.

Chapter 5

The Arc of the Trapeze

The design review with Cummins produced more technical
clarity and a better relationship. The design, however. . . .

—*Neill Lane reporting on the 5K, February 1990*

THE STIRLING ENGINE that Neill and Abhijit and the
other members of the 5K team undertake, in the fall
of 1990, to design bears some vague resemblance to
other Sunpower engine designs. It does not, however, resemble anything else in the world. Sunpower has a rubber stamp that is supposed
to be inked across every design drawing that leaves the place, warning
in harsh terms that the drawing contains proprietary information and
is the sole property of Sunpower, Inc. This is a classic example of
William Beale's hope triumphing over experience, in that most outsiders—even engineers—cannot make heads or tails of a Sunpower drawing, and it is not because they are not elegantly drawn. It is because
most people—even most engineers—live in a rotational, internal-combustion, fossil-fueled world, a world that shapes one's expectations of
what engines look like. No internal-combustion engine looks like a
Stirling engine; few Stirling engines look much like the solar 5K as it
begins to take shape within the minds of the engineers and Todd Cale's
computer drafting machine.

As might seem appropriate for an engine that doesn't resemble
much else in the real world, the business plan that begat the 5K project
doesn't much either. In its simplest form, the 5K program at Sunpower
is this: design, fabricate, and test a Stirling engine capable of producing
5,000 watts of 240-volt, 60-hertz alternating current when powered
by the sun's heat; do it in conjunction with a small company that
specializes in heat transfer, with a small company that specializes in solar

concentrator dishes, with a small company that specializes in electronic control systems, and, above all, with Cummins Power Generation, Inc., a subsidiary of the largest manufacturer of diesel engines in the world, the Cummins Engine Company, of Columbus, Indiana. And do it for, in round numbers, a million dollars.

What CPG engaged to do was sponsor the research, development, and construction of what Cummins thought was a device with tremendous potential: a solar-powered engine that generated alternating current, a machine that would hook into the nation's electrical grid. One machine—one ultimate machine, after all the R & D—would produce, say, 9 kilowatts, and depending on its design and its potential configuration (possibly in pairs or threes or fours on one solar dish), might produce 25 kilowatts, net, of electrical power. Ten solar dishes each producing 25 kilowatts would then produce 250 kilowatts; a hundred dishes, 2,500 kilowatts, or 2.5 megawatts, which is getting up into the real power range—a rule of thumb for calculating the electrical needs of a city is about a kilowatt per person on the average, so a hundred-dish, 2.5-megawatt installation would cover the electricity needs, in broad numbers, of twenty-five hundred people. A thousand dishes could serve twenty-five thousand people, and so on. The engines might be expensive in the short term, but the *fuel* is free: in an era where conventional fuels are more and more costly and more and more politically volatile, and under more and more scrutiny for their environmental effects, free nonpolluting fuel looks very good. So CPG signed on with Sunpower to do some R & D.

In the real world, where companies like Cummins are used to operating, it is impossible to tell what is altruism and what is good business; in the Sunpower world, altruism *is* good business. In this case, though, if the project worked out, good business for Sunpower could also be good business for CPG, which doesn't happen all that often, pastoral pictures of caribou grazing along the Trans-Alaska pipeline notwithstanding. CPG was interested in Stirling technology because its CEO, Jerome Davis, and two of its senior engineers, Rocky Kubo and John Bean, thought the technology was viable and potentially profitable. Sunpower was interested because it thought that this might be the big chance to make a *commercially* viable machine, one that would eventually be sold in the thousands or tens of thousands of units, which would not only establish the technology in a big way but provide Sunpower with the resources to design and create other machines (like William's small engine).

What Cummins contracted for was not a pig in a poke, exactly.

The 5K's immediate predecessor in hardware was an engine called the SHARP (a tortuous acronym for Stirling Helium Advanced Reliable Prototype), originally designed for a United States Army competition to develop a quiet and reliable electrical generator (a competition that Sunpower did not win). The SHARP's gross design output was two kilowatts. The first step, then, in the 5K program was a "proof of concept" trial for CPG—CPG, it cannot be stressed too often, is a company doing business in the real world, so it requires trials and demonstration projects for any new technology it considers in an effort to keep up-front costs low during the time when a potential project is being evaluated for a full-scale program.

The proof of concept in this case involved pairing two SHARP engines on one solar dish, with the goal of producing a net output of three kilowatts of electrical power from the sun. (The SHARP Army project also involved pairing two SHARP engines, although they were powered by a natural-gas burner, rather than by the sun.)

The rationale for doing a proof of concept with the paired SHARP engines was twofold. First, it would provide an opportunity to test the other components necessary to running a Stirling engine on sun: the solar concentrator dish, the control system that would tilt the dish as it tracked the sun across the sky, the "heat pipe" upon which the sun's rays would be focused and which would thus channel the heat the engines need to run. Second, the proof-of-concept trial would be accomplished using—by Stirling-engine standards, anyway—a relatively proven engine design, which Cummins and Sunpower thought would help put things on the fast track. Two of the engineers on the 5K, Abhijit Karandikar and Joerg Seume, were part of the SHARP proof-of-concept team, and Neill Lane had worked on the original SHARP project, back when it was part of the Army generator program. So, on paper at least, the SHARP was for Sunpower an excellent starting point for an ultimate machine.

The initial plan discussed by CPG and Sunpower called for an ultimate machine producing eight kilowatts, which before the testing of the paired SHARP engines was possibly going to be a pair of four-kilowatt engines. After the testing of the two SHARPs, the decision was made, for several reasons that seemed compelling at the time, to make a single eight-kilowatt engine. Part of the rationale for this was the ongoing experience—what turned out to be a not altogether happy one—that Sunpower was having with NASA, in which Neill and Jarlath McEntee designed a twenty-five-kilowatt Stirling engine.

Early on in the negotiations with CPG, information gathered during the NASA design competition (which was initially considered to support the idea of a very large engine) ended up discouraging CPG and Sunpower from pursuing a very large engine. And then, in relatively short order, and for reasons that again seemed compelling at the time, the eight-kilowatt engine became a five-kilowatt engine; then the fundamental architecture of the engine was changed; then, after three months and $350,000 worth of design work, its fundamental architecture was changed again. What had started out for Sunpower as looking like a fairly straightforward modification of a design with which it was fairly pleased, began to turn into what engineers, especially Stirling engineers, fear the most: a clean sheet of paper.

In the larger scheme of things—the long view that successful companies are always striving to preserve—the sorts of changes that turned the project from a modification of a proven design to a completely new design are not only predictable but expected: no one believes that the first idea to arise from a discussion will be implemented in the exact form of the original conception. That's being realistic. On the other hand, no one believes that the entire conception of a project will change so many times, with so many ramifications. That would be pessimistic. But it can be frustrating all the same, because it's only after something doesn't work out that it becomes apparent that another choice might have been made.

The Cummins 5K project marked the closest that Sunpower had ever come to a certain kind of machine, one intended to be a major player in the world energy game; because of this, the scrutiny—and the pressure—were intense. When the rules of the game changed, Sunpower, and the design, had to change. Otherwise it couldn't play. In Sunpower's long and difficult history, it had ended up being dealt out of a dozen different games for a dozen different reasons, ranging from bad luck to bad choices to bad economic times to simply bad karma. But the Cummins project was so close to the idea of a real ultimate machine that both Sunpower and Cummins forced themselves back to the table time and time again, for one more hand. In the short term, the million dollars for initial R & D was important to Sunpower's financial health; if the program was successful, the million would turn into several more millions for more prototypes; beyond that, if the machine actually made it from the drawing board to the free market, Sunpower would get a royalty for each and every machine that Cummins manufactured and sold, providing money for more R & D on other projects. It would thus operate just like a company in the real world.

So the stakes were high for the 5K team and for the company. This was why Neill and Abhijit seemed, especially at the beginning of the project, to spend as much time on management as on design, why they lobbied for technicians as well as engineers, for test rigs as well as engine parts: when problems emerged, they wanted to be on top of them, anticipate them, rather than being stupefied by them. They knew, or thought they knew, how to do everything they had to do. They needed, however, to also do it, which is not quite the same thing.

In all fields, there is something that can be called theory and something else that might be called practice; in literary studies there are critics (theory) and poets (practice); in high finance there are economists (theory) and bankers (practice). In the field of Stirling engines, there are thermodynamicists (theory) and design engineers (practice). As in all other fields, there is room for overlap. Just as there are critics who write novels and poets who write criticism, there are thermodynamicists who get their hands dirty and design engineers who can do equations; as in other fields, there are some few who do both equally well, as well as many who do one much better than the other.

Most people and, indeed, most companies specialize, though, in one or the other. There are Stirling-engine research companies in the United States that survive solely on doing feasibility studies and grant-funded analyses and simulations, activities where the engineers get their hands dirty only when they change the ribbon on the computer printer. (This is what drove Neill Lane to near-madness on the NASA project: it was all on paper, and looked to stay on paper for the foreseeable future; he once said it resembled psychodrama more than it did engineering.) If Sunpower were interested only in demonstrating how marvelously Stirling engines work in a computer simulation, it would have no need for a lab, or test cells, or a machine shop. Overhead would be much lower, because changing a number in a computer program is much easier than refabricating a part, or a dozen parts, in order to make a subtle change.

But Sunpower's dream—and the 5K team's job—is to *make* Stirling engines, and to make Stirling engines, you need a lab and test cells and a machine shop and tools and big pieces of metal. And people with the skills and talents to turn big pieces of metal, in machine shop, lab, and test cell, into working hardware. Dreams are easy—even big dreams. Hardware, however, is *hard*.

So to make an engine design into an engine requires three components, all in large amounts. The first is theory; the second is practical skill. The third is money, which buys the first two. Around Sunpower, there is no shortage of the first, plenty of the second, and never enough of the third, in large part because the first two are so expensive. They are so expensive because everything must be envisioned, designed, and fabricated virtually from scratch: the off-the-shelf parts in the 5K engine could be carried around in a briefcase—the bolts that hold it together, the plugs that connect its electronics, the rubber gasket–like O-rings that seal its joints. Virtually everything else is handmade, one of a kind, having made the transition from theoretical function to imagined design, to design on paper, to fabrication.

When the design of a machine changes early in its conception, the changes might involve some computer calculations or changes in a drawing. Later in the process, a part or a part of a part might be redesigned or reconceived to make it smaller, or stronger, or lighter; if the parts that it fits into, or the parts that fit inside it, haven't been made, these sorts of changes are also relatively easy to implement. At some point, however, a critical number of the parts are fabricated, and all design decisions and modifications become subordinate to the parts already done. Sometimes this is a case of money—it would cost too much to refabricate. More often, it's a case of time: at some point, there is no turning back, no starting over. The only way out is forward, so the design process, if it is thought of at the beginning as consisting of a limitless number of possible routes, over time narrows further and further until there are very few options at all. When a program reaches such a point, its course is to a large degree set in stone. The time to explore possibilities is earlier, rather than later, in the process.

The 5K team knows this, of course, both in theory and in practice. Neill Lane, in fact, takes it as a point of pride that the 5K will be the most smoothly managed and implemented project in Sunpower's history. It will not become a Sunpower legend for any of the wrong reasons—not working, not being documented, not being understood. If the 5K will be a Sunpower legend at all, it will be because it has been done by the book, the book that Neill and Abhijit will write as they go.

Neill likes to make a point about the right way and the wrong way to do things by referring to several engines from the SHARP program, which, while working fine, were each one-of-a-kind objects in themselves, with slightly different designs for each. "*None* of the SHARPs had interchangeable parts," Neill will say, with some exaggeration. "There was *nothing* in them that was made the same way *twice*." By contrast, Neill plans to ship a complete set of drawings for the 5K along with the engine, when shipping time comes, a set of drawings that will not only reflect accurately the design and fabrication of the engine (one coup) but also be usable, start to finish, in making *another* engine (quite another coup). So immediately Neill gains a reputation at Sunpower of being positively *obsessive,* a reputation he despises. When he and Abhijit make the "two-initial rule" for the drawings, this reputation is enhanced, and even Neill admits it is to some degree deserved: the two-initial rule states that no design drawing may leave the drafting room without *two* engineers on the project understanding the drawing in its entirety. If Neill Lane makes a change in a drawing—adjusts a tolerance, moves a bolt circle, makes a part thinner—he has to find another engineer on the project to explain the change to, and they both must put their initials on the drawing.

The two-initial rule came about from Neill's and Abhijit's understanding of how things at Sunpower had often gone wrong in the past. One of the profound differences between tinkering and engineering is that an engineer's work is not considered done until it can be replicated by someone else according to standard practices. A machine that no one but the inventor or designer understands is sculpture, not engineering. And despite having an impressive roster of engineers on staff throughout the company's history, Sunpower has made its share of sculpture—engines modified and reconfigured and improved so many times during fabrication that the final product (even if it ran wonderfully) would be reproducible only if it were disassembled and each part drawn after the fact, instead of having been made from the specifications in a drawing. Neill's occasionally dark sense of humor also causes him to observe that, in the past, "there have been designs here where if the engineer in charge got hit by a car, the company would have to start all over again from scratch"—he once said that he didn't want his last words in some ambulance to be "Ehhh, tell Abhijit that I made this change in the drawing."

If Stirling-engine design were a more exact science, this would be less of an issue, but the machines are *so* idiosyncratic, so much the product of many hundreds of small decisions, that retracing one's steps—even when one knows exactly what's going on—can sometimes take hours. And sometimes, it's impossible. (Robi Unger, an engineer on Sunpower's Stirling-cooler projects, always answers questions about why something works by saying that he doesn't know. He's kidding only about 75 percent of the time.) So Neill's fears, though macabre— "Drive carefully!" he once said to Abhijit when he left for a weekend trip, "or we'll go bankrupt!"—are not unreasonable.

Auto accidents notwithstanding, the two-initial rule illustrates the complexity of invention. One engineer *might* be able to design the 5K, but it would take many times longer than the time allotted; one engineer *might* be able to do all of the equations necessary to ensure that the engine would not only work but do what it was supposed to do, yet this would be literally thousands of equations: stress analyses, mass calculations, materials analysis and selection, thermodynamic simulations, calculations of vibration, of temperature, of pressure, of frequency. One engineer *might* be able to do this, but the world would grow old waiting. So there is division of labor, and division of labor increases the chances that the left hand will not know what the right hand is doing.

How this can happen is easy to understand and difficult to solve. More than one engineer always has a role in the design process: on the 5K, besides Neill (as senior design engineer) and Abhijit (as project manager), there are Barbara Fleck and Joerg Seume on the team as design engineers, Robert Redlich (as a consultant for the alternator design), and, of course, William (as technical director). The work is divvied up according to time and expertise; Abhijit designed the band clamp that held the pressure vessel closed on the SHARP project, so he designs the band clamp for the 5K; Joerg Seume's Ph.D. dissertation at the University of Minnesota analyzed heat transfer and fluid mechanics, so Joerg has a hand in the design of the 5K's heat exchangers. Joerg's background in tribology—the science of friction and wear— made him the likely person to design bearing surfaces as well. And so on. This is Sunpower's biggest treasure: engineers who are very, very good at the arcana of Stirling-engine design, who are good at math (many engineers, it is surprising to learn, are not), who are skilled and experienced in ways that only a handful of engineers around the world are skilled. But with wisdom comes responsibility, at least as far as Neill and Abhijit are concerned.

Where it might seem enough that an engineer on the project knows how to answer a design question, for the good of the program it is not: the engineer must also be able to explain the reasoning to another engineer on the program. When everything is clear, they initial the drawing. When Neill and Abhijit dreamed this up, they envisioned tweedy conversations over coffee—miniseminars, really, in which one engineer explains how something should be done to another engineer, who then understands it. They would hold the line on this.

And they did, until the day when, under pressures of time and circumstance, one of the engineers shoved a drawing at another engineer and said, "Hurry up and initial this." Depending on who is doing the ribbing, and who is being ribbed, the shover and the signer are identified differently each time the stories are told, but the picture is familiar to each engineer, because all have been signers, and all have been shovers. One afternoon, Neill Lane saw Barbara Fleck, coat on, on her way out. "Barbara!" he called. "Come in here and initial this drawing, please."

Barbara comes in. "What is it?"

"Don't worry about what it is; just initial it. I want to get the part into the shop."

Another day, Barbara Fleck had a drawing for a small part used in a springing test for the engine. She found Neill Lane having a cigarette. "Here, sign this," she said.

"What is it?"

"Don't worry about what it is; just sign it."

And so things go.

To understand what the 5K team is doing—beyond shoving drawings at one another to be initialed—is to understand what it means to be a design engineer. The final product that the 5K team hopes to make requires expertise not only in Stirling engines but in electrical engineering, mechanical engineering, dynamics, statics, physics, mathematics, production engineering, computer science, electronics, drafting, and materials science. Also, fabrication techniques, metallurgy, tribology, heat transfer, fluid mechanics, flow mechanics, chemistry, conduction, convection, diffusion, and creep. Also economics, interpersonal communication, management theory, finance, accounting, history, political science, and, to no small degree, psychology. A design engineer might aspire to being truly expert in one or two or possibly three of these areas; the engineers on the 5K are, in toto, expert in something like half of these fields, and other engineers and consultants at Sunpower cover perhaps another third. In the areas left over, they tend to wing it. But even in winging it, they need a rationale that everyone can see; everyone might not agree, but everyone must understand.

Neill Lane is an intensely design-oriented design engineer, an honorific that his training has taught him is special. He is, in fact, one of three University of the Witwatersrand-trained engineers at Sunpower, the other two being David Berchowitz and one of Neill's own former students, Nick van der Walt. Whereas many American engineers might have a design course as, say, a senior-year elective, the program at Wits is in essence a four-year course in design engineering. It gives Neill Lane a different perspective on how to approach, understand, and try to solve design problems, such as how to make the 5K go.

Even the *name* of the 5K program bears Neill's stamp. By Sunpower convention (and the entirely natural tendency of humans to be optimistic), the 5K would actually have been christened the 8K while it was still simply a string of thermodynamic equations, eight kilowatts being the computer simulation's prediction of the power the engine would produce. But the computer simulation is always overly optimistic, and is automatically reduced in the minds of the engineers by anywhere from 10 percent to as much as a quarter (the larger the machine, the larger the reduction). Always. And everyone knows this. And when the power requirements of the machine itself are taken into account—how much electricity the engine, its controls, its solar-tracking mechanism, and so on will use—the grossly optimistic gross figure is reduced further still. But the Sunpower convention has long been to call engines by their design output, so the 5K might well have been called the 8K from the word go.

Neill, however, was aware of the politics and the sociology of naming; he knew the engine wasn't an eight-kilowatt engine (the working goal, before the necessary reductions for its own power consumption, was closer to six kilowatts), and he didn't want to be the designer of an eight-kilowatt engine that produced "only" five kilowatts. So he insisted on calling a spade a spade, and the 5K was the 5K.

At thirty-two years of age, he is young by most engineering standards for a position, even at a small and idiosyncratic company, of such responsibility; he was young for his profession, too, when he taught as a junior lecturer at Wits. As the 5K team is assembled, though, he looks comparatively older: they are all young. Abhijit Karandikar perhaps looks the youngest.

When Abhijit first joined Sunpower, in 1987, he was thrown, without much formal introduction, into a David Berchowitz–managed engine project, known around Sunpower as "the 3KW Air." (Rather than using helium as a working fluid, this engine was designed to use air, hence the name.) Dave Shade, Sunpower's most-senior test engineer, recalls how he was deep into debugging the 3KW Air under a vicious deadline when Abhijit was led to the test cell, introduced as "Beej" in a noisy room, and left there, ostensibly to get right to work on an engine he had never seen, for a project he knew nothing about, with people whose names he didn't know. (Introductions at Sunpower are extremely casual most of the time.) For a while, Shade thought that Abhijit was some sort of visitor; later, he decided he was some sort of consultant, or perhaps a student. Perhaps he was a new technician, or an undergrad from the university? Shade tested this theory by sending Abhijit to fetch tools. Abhijit, eager to contribute in any way he could while he got his sea legs at Sunpower, fetched tools.

Fresh from two years in the West Indies, where he worked for the Peace Corps, Abhijit also had a habit of greeting people by saying, *"Hola, Paco,"* which puzzled Shade in the extreme; the new technician/visitor/undergrad/stranger looked as though he might be from India or Asia, but he sounded as if he spoke Spanish with a Canadian accent. Shade tried to be as friendly as he could under the circumstances, calling Abhijit what he thought he had heard when Beej was introduced as Beej, which was "Habib," all the while waiting for the new engineer from California who was supposed to help on the system coordination for the 3KW Air. (Abhijit, who was born in England and raised in Canada by his Indian parents, went to school at Cal-Berkeley, and engineers tend to be thought of as being from where they went to school, as though they had been born in engineering class.) This went on for several days.

For a while, when Beej heard Shade talking to someone named Habib, Abhijit thought he was talking to someone else, until Shade called him Habib when only the two of them were in the room. Then Abhijit tried to figure out why Shade was calling him Habib. Was it an ethnic slur? And why was he always being sent to fetch tools? Wasn't Beej supposed to be doing *design* work on the 3KW Air? Finally, Beej couldn't stand it any more. "Why are you calling me Habib? My name isn't *Habib*," he said sternly; "it's Abhijit Karandikar, and I'm not a go-fer, I'm an engineer who's supposed to be managing *you*."

"Are you the guy from California?" Shade asked, putting two and two together.

"Yes, I am," said Beej, wondering whether Shade was prejudiced against *Californians*, too.

"Oh," said Dave Shade. "Sorry." Nothing but kind words have passed between the two of them since.

Beej *does* look young; he could still pass for an undergraduate on a college campus, but looks are deceiving, because he has the most "adult" job on the 5K project, that of management, system integration, and "sponsor interface," which means he is the one who keeps CPG up to date on the team's progress and keeps the team up to date on CPG's thinking. He is the one who relays bad news from Sunpower to Cummins (when there has been a complication or a failure); he relays bad news from Cummins to Sunpower (when there is a disagreement on strategy, policy, or progress); when Neill, in a fit of anger, suggests that Cummins perform what would be an anatomical impossibility even if it weren't a corporation, Beej is the one who tells Cummins that Neill "has a concern." When things are really tense, he might suggest that there are "strong feelings." Abhijit is a consummate diplomat, a negotiator, a conciliator, a counselor. As project manager, he is responsible for everyone on the team having just enough to do, but making sure that what needs to be done gets done. At its largest, the 5K team is more than a dozen strong, including technicians, machinists, draftspersons, and engineers, and Beej takes it upon himself to know what everyone is doing, when each might be done, and what will be done next. For a project that consists of many hundreds of parts, each of which must be sent on a path from design and specification through raw-material procurement, delivery, fabrication, quality control, and assembly, it is good that this is the sort of thing upon which Abhijit thrives, especially since this sort of thing drives Neill Lane to distraction. Neill's style of management when under pressure is to give an order—politely and with a smile, but an order just the same; Abhijit turns every task into teamwork. The sheer number of parts, subassemblies, assemblies, rigs, jigs, processes, and procedures that the project requires is incalculable. It is as though one were set upon the task of designing and fabricating an automobile from a truckload of pieces of metal—and without having an automobile in the driveway to use as a model or template. Many of the tasks draw on Sunpower's previous experience; many others are done from scratch.

But someone (and by the 5K team's rules, more than one) must have a good grasp at all times of how *everything* relates, how the alternator affects the engine, how the electronics affect the engine, how the electronics affect the alternator, how the engine affects the electronics, how a change that might be trivial to the design and function of, say, the piston gas spring might have a tremendous impact on nine other things. A good design team for a Stirling-engine program resembles a good design team for something like a woven tapestry, everyone working on some particular part, but always as a part of the whole.

What makes engine design particularly complicated is the relationship between dynamics and thermodynamics—between the characteristics of the moving parts in relation one to another in the system, and the characteristics of the flow of energy through the system. It is as though one were trying to illustrate an animated cartoon in tapestry: the whole thing has to be rewoven every time a change is made. Likewise, a subtle change in the thermodynamics can have dramatic implications in the mechanical dynamics, and vice versa, in large part because the dynamics and the thermodynamics of a Stirling engine are unique—an overused word, but in this case, absolutely accurate: a free-piston

Stirling engine, when running, requires a leap of faith to understand, the the more one knows afterward, the more—rather than less—astonishing it seems. And it is like nothing else.

The unique thermodynamics of a Stirling engine result from the way heat moves through the engine. There is, of course, a piston, which does the work. And there are two "heat exchangers," one hot and one cold, which are the places where energy in the form of heat enters and leaves the system, respectively, and which would be a part of any heat engine. But other engines have nothing like what Robert Stirling, that thrifty Scot, called the "economiser," and what contemporary Stirling designers call a regenerator.

The regenerator is a space filled with anything that will absorb heat when hot helium passes through from the heat source of the engine, and give off that heat when cool helium passes through it on its way back to the heat source. The key to the regenerator is "surface area"— the more surface area the regenerator has, the more places it has to absorb heat. A regenerator could be a solid chunk of iron (in early engines, and in some very simple, very low-power engines, the regenerator is merely an "annulus," the engineer's way of saying empty space), because a chunk of iron (or the air in an empty space) will absorb heat when heated and give it off when cooled. But a chunk of solid iron has only the surface area of its height and width and depth; cut a series of slices out of a chunk of iron, and you gain surface area; you make fins. Look at an old-fashioned cast-iron radiator, which is a series of tubes, tubes with fins, rather than a solid chunk. The more surface area something has, the more places it has to absorb, or give off, heat. (Another example of surface area is the lining of the stomach, with its many series of folds; the stomach evolved this way to provide a large surface area for the absorption of nutrients, but the principle is the same.)

So a regenerator wants lots of surface area: many thin layers of stainless steel foil, for instance, or "random" wire that looks like glorified steel wool, although it is made out of stainless steel of known metallurgical properties, rather than out of something like junk cars, which ordinary steel wool can be.

The regenerator is important thermodynamically because it stores heat during the Stirling cycle. But for the regenerator to work, there must be a good way to move the helium through it, which is the displacer's job; the displacer forces helium through the regenerator and into the space between the piston and the displacer: the helium, when it's hot, leaves some of its heat in the regenerator on its way through in one stage of the engine's cycle; when the helium is cooled, it is forced back through the regenerator and takes some of the heat stored there, and thus heats up again. The higher the pressure of the helium in the cylinder, the more energy it contains—although a thermodynamicist would hate to think of it in quite this way, more pressure means the gas moves "faster"; a nozzle on a garden hose restricts the passage through which water can flow, and thus increases the pressure, so water comes out of a nozzle with much more force than it does out of the faucet. Double the pressure, and double the speed at which the gas in a Stirling engine moves from hot space, to cold space and back again. These are the components that work together in such a way as to create energy flow through the system: the thermodynamics.

The mechanical dynamics, on the other hand, are the moving parts. Focusing for a moment on the movement of the mechanical parts inside the sealed pressure vessel is one way to appreciate just what happens inside an FPSE. Inside the pressure vessel is the concave, disk-like piston, which in the 5K is about the size and shape of a soup plate. The piston fits very closely inside its cylinder, a close fit enhanced by rings around its outer edge that make for a very tight seal, but the piston is not attached to anything in this world; all that moves it when the engine is running is the action of the helium as it expands and is compressed as it moves through the heat exchangers and the regenerator. Part of the piston assembly must be some sort of arrangement for getting the work out of the engine; if the work to be done is the generation of electricity, then the piston is not attached, really, to anything in this regard, either: in an alternator like the one the 5K team will need for the Cummins project, the piston is connected to a ring of magnets that pass between the inner and outer windings of the alternator, but the magnets pass through a gap, rather than being attached to the windings.

When a Stirling engine is running, the piston moves back and forth, pushed in one direction by the expanding helium, and bouncing back when it compresses the helium in the other direction. (In a pre-Beale crank engine, the piston would have to be attached to some sort of mechanical linkage.) But in the FPSE the piston assembly *rests* on nothing; it is captured in space, though not tethered.

And yet it moves back and forth at a certain frequency, a frequency calculated by the engine's designer on the basis of the characteristics of the piston and its related parts, the pressure of the helium, and a long list of dynamic and thermodynamic factors. The result of these calculations will be a "natural frequency" at which the piston "resonates," or moves back and forth at its maximum amplitude, and can be thought of as similar to the way a basketball bounces in the hands of a skilled player: the ball seems to be held by some force not only in the path between fingers and floor but at an amplitude (how far from the floor the ball bounces) and frequency (how many times per minute the ball bounces) that requires very little adjustment by the dribbler. A good basketball player can bounce a ball, literally, on a dime, at the same height, and at the same frequency, for a long time; the ball bounces from the floor to the fingertips and back again, over and over, in a rhythm.

Yet the ball itself is "free" in the same way the piston in a FPSE is free: only a balance among the many forces acting upon it keeps it moving at a predictable frequency and amplitude; change the forces, and the frequency or the amplitude or the path changes.

A good basketball player can easily control a ball bouncing perhaps once per second, sixty dribbles per minute. In the case of the 5K, the target frequency is sixty oscillations per second—60 Hz—3,600 trips back and forth per minute, 216,000 per hour, 1.8 million cycles in an eight-hour run. In an 8-kilowatt engine, this would produce 64 kilo-watt-hours, enough power to light eighty 100-watt light bulbs during a workday. Another way of thinking of this is that 1.8 million cycles would produce about three dollars worth of electricity at typical rates. This is the reality that the 5K team lives with: the ultimate machine must not only work but work so well that 1.8 million cycles is a tiny milestone in a real-world sense, because if the 5K costs a million dollars to develop, it has to have the potential of producing many millions of kilowatts if it is ever to pay for itself.

But a piston by itself wouldn't produce a single watt, because a piston by itself wouldn't resonate in a thermodynamic system: heat pours in at one end, expanding the helium and moving the piston back, but the comparatively weak forces that accumulate on the nether end of the piston as the helium there is compressed would never counteract the force generated by the introduction of heat. The piston would be pushed back and stay back.

So there is the displacer. The displacer resembles the piston in that it too is captured inside a cylinder, but because its role is to displace helium through the heat exchangers, it is shaped like a can, say, a can of green beans. The displacer's volume takes up space. A slender rod, about the diameter of a walking stick or cane, runs from the displacer, through the piston, to another cylinder, where it is connected to a round puck. This is the displacer spring: when the displacer moves back, pushed by the expanding helium in its cylinder, the puck moves back as well, compressing the helium behind it; there, the helium pushes back and acts as a spring—a spring of compressed gas. This spring pushes the displacer forward.

The displacer, like the piston, is designed to operate at a particular frequency, to resonate, as the piston does, at 60 Hz. But if the piston and the displacer simply moved in phase with each other—back and forth sixty times per second in tandem—they might as well be attached, and a piston and displacer that are attached one to the other would behave much like a piston alone.

No, the displacer's frequency must be in exquisite synchronization with the piston, but it must be somewhat out of phase as well. It must lead the piston. It must make the first move, moving at the same frequency as the piston, but in such a way that it is always moving forward before the piston moves forward, always moving back before the piston moves back.

Too much of a lead, however, wouldn't be any good either. If the displacer were moving at all times in opposition to the piston—in when the piston was out, and out when the piston was in—they would be perfectly out of phase: and they would collide in the middle of their respective strokes, because the piston and the displacer must both occupy the same place within the cylinder, but at different times.

Trapeze artists know this, or learn it. Two trapeze artists swinging in the same plane will collide unless they are swinging in a particular phase relationship to each other: if the two trapezists start at the same time and move in phase, they will swing back and forth like windshield wipers on a car, never banging into one another (which is good), but never being in the right place for one to leap into the other's arms (which is not good, or at least not very interesting to circus patrons). Likewise, if the two trapeze artists are exactly out of phase, they will simply smash into each other, since both trapezes need to pass through the same bit of space in order to complete their arcs. This is not very interesting either, to watch or do.

Instead, trapeze artists swing in a phase relationship in which one trapeze leads the other, passing through the point of collision first, moments before the other trapeze. The first one through lets go, and falls into the arms of the other, who is moving up close behind. This sort of phase relationship between two moving objects can be sketched out using lines called vectors, which engineers use to describe the position of an object in terms of its magnitude—how far it travels—*and* its direction. A vector diagram of a trapeze act would show one artist leading the other: one is first in a particular place at a particular time in a particular direction, and the other occupies that place at a slightly different time. And when there is a spectacular miss, and one acrobat tumbles into the net, it will be because, as sometimes happens in a Stirling engine, the dynamics weren't quite right—not enough velocity on one trapeze, too much on the other, a change in body position (which would change wind resistance and aerodynamics), a whole host of factors. The margin of error is really quite small.

Once one starts looking for them, phase relationships are quite common, although few are as elegantly cyclic as a trapeze act. (Neill Lane's analogy for a cyclic relationship in which phase is important is the act of sexual intercourse, although he hasn't gone so far as to draw a vector diagram or do the calculations.) In a football game, the quarterback throws the ball not *to* a receiver but to where *the receiver will be* when the ball gets there. He "leads" the receiver, because the receiver, as well as the ball, is moving—a receiver and the football must occupy the same place at the same time if the pass is to be complete, but because both are moving, the quarterback adjusts his pass, the receiver his pattern, in order for the two to arrive together. In a baseball game, double plays are turned because the shortstop throws the ball to second base—not to the second baseman, who is still on his way to cover. The second baseman and the ball arrive at the same time, and the second baseman pivots and throws to first, where the first baseman will be moving toward the bag, but might not be there yet.

Windshield wipers, trapeze artists, second basemen, two men driving a railroad spike, members of a relay team passing the baton, all require a particular phase relationship and could be drawn in a vector diagram.

The phase relationship between piston and displacer that works best in a Stirling engine is one in which the displacer leads the piston by something like one-third of the length of the stroke; that is, the displacer is always one-third of the way forward when the piston begins to move forward, and one-third of the way back by the time the piston begins to move back. This phase relationship works because of what is implied by the relative positions of piston and displacer: when they move, they are in effect changing the volume of the cylinder in which they ride, because the piston and the displacer are the two ends, bottom and top, of the cylinder. The bottom (or back) of the displacer is actually the "top" of the compression space, and the piston is the bottom of the compression space. Similarly, the top (or front) of the displacer is actually the "bottom" of the expansion space. This is a good moment to think again of ice cream and the push-up: the bottom of the ice cream is like the bottom of the displacer, and the space you make by pushing the ice cream up, and then drawing the cardboard piston back, is analogous to the compression space in a Stirling engine.

But describing the phase relationship in such language as "one-third" is obviously imprecise when it comes to actually designing an engine, so in describing the location of the piston relative to that of the displacer, Stirling engineers use a convention borrowed from vector diagrams. The relative positions are defined by the geometric angle their vectors form, and thus are spoken of in "degrees," as one would speak of, say, a "90-degree angle" when describing the intersection of perpendicular lines. In Stirling phraseology, then, the "phase angle" describes the relative positions of the piston and displacer: a phase angle of zero degrees means the piston and displacer are in essence traveling together; a phase angle of 180 degrees would describe the piston and displacer banging into each other in the center of the cylinder from the extremes of their amplitudes (the image of two trains on the same track traveling in opposite directions and converging on a railway station is an arresting way to remember this). Phase angles between zero and 180 degrees describe conditions where the displacer is leading the piston; those from 181 to 359 degrees would be phase angles where the piston is leading the displacer (360 being indistinguishable from zero, by this convention, in that if both are traveling together, neither can be leading or lagging). Thus, rather than saying the displacer leads the piston by one-third, one describes a phase angle of 60 degrees (or one-third of 180, the range in which displacer leads piston).

All of this might be only interesting, rather than crucial, if it weren't the phase relationship between the piston and displacer that made Stirling engines go. If one imagines the path of the helium inside the pressure vessel, it's possible to understand why this is so. In order for the engine to run—that is, to move cyclically and do work—helium must be heated and, thus, expanded in the end of the cylinder where heat pours in from the heat source, say, a gas-fired burner. This is the expansion space. While the helium is at its hottest and most expansive, it expands in the cylinder, driving the piston out and causing the piston to do work. But the displacer is leading the piston, so the helium is also pushed by the displacer through the regenerator, where much of its heat is stored in the thin metal foil or random wire packed there; the helium is further cooled by then passing through the cooler, where a jacket of circulating water (similar to an automobile radiator) extracts more heat, just in time to arrive in the compression space, where the piston compresses it, because the piston is lagging the displacer, and because it is "bouncing" back from the opposite end of the cylinder. The helium, because relatively cool, is easy to compress, and this forces the helium back through the cycle—into the cooler first, then into the regenerator, where it picks up some of the stored heat and expands, driving the piston out again in a stroke that does work. When a Stirling engine is running, it is cycling like this sixty times a second; the only moving parts are the displacer and the piston as they bounce back and forth, alternately compressing and expanding the helium, forcing it through the heat exchangers. If the pressure of the gas is right, and if heat is continually added through the heater head, and heat continually rejected through the cooler, and work continually taken out via the alternator, a Stirling engine will run like this for as long as its contact surfaces—the rings around the piston and displacer—keep a tight seal. An engine designed with "clearance" seals (which is just a very narrow gap, rather than an actual contact surface) will run, theoretically at least, forever.

It is an exhilarating concept, in large part because it seems to improve upon what are more and more appearing to be obvious disadvantages of internal-combustion engines. For one, it can be run on any fuel—any heat source will do. For another, a Stirling engine is sealed into its pressure vessel and needs no lubrication, no maintenance, no oil change. (A solar dish system requires regular maintenance for the dish itself, the sun-tracking system, the controls; this would be done at night.) And it makes electricity, for which a distribution system is already in place and which is already routinely shipped hundreds of miles as a matter of practice. And for all its hundreds of parts, and thousands of calculations that are necessary to invent, perfect, and bring the first engine to life, it is a relatively simple mechanical device.

That's the theory. The practice occurs in the drafting room, the test cell, the machine shop, and the lab. The 5K team starts out to build the first 5K engine.

Chapter 6

Turning Money into Hardware

Steady progress. Rapid learning curve on handling magnet paddles; the paddle broke within a couple of hours of its arrival, but it *did* break in its special booth.

—*Neill Lane reporting on the 5K, July 1990*

I T IS COLD in the 5K test cell on this gray December afternoon. The floor, which is poured concrete, seems to *radiate* chilled air, in apparent violation of all the laws of thermodynamics. The people in the room stand on one foot and then the other, in an unconscious sway, as though lifting one's foot from the concrete for a moment would take away some of the chill.

Dave Shade is a one-man flurry of activity. He wears a dirty lab coat that was once a color Sunpower employees refer to as "Smurf blue," and walks purposefully in and out of the test cell, on one trip carrying a tool, on the next carrying a new rubber gasket, on the next a roll of black electrician's tape, as he assembles part of what will someday be a Stirling engine. Today it is just a "test rig": part engine, part monitoring devices, part test equipment. If the test is successful, the team will go on to make the rest of the parts necessary for the test rig to become an engine.

The test cell is really two rooms; the outer area is a control area, with a computer terminal, a bank of oscilloscopes, a rack of switches and dials and gauges. A ceiling fan whirs, ostensibly pushing heat from the space heater down into the room, but in practice adding another cool breeze to a room that seems full of cool breezes. The building in which Sunpower is located was never particularly frilly; it was a machine shop, a warehouse, a laboratory for a university engineering department, a place where jukeboxes were once repaired. The walls are concrete block; the doors and windows, utilitarian and untrimmed. The ceiling in the test cell is extruded sheet metal; the rafters are bare. On the wall of the control area hangs an assembly drawing of the test rig, and a schematic for the electronics that will control it. Printed on blueprint paper the size of movie posters, they are the only form of decoration in the room, and not much decoration at that. The room is all business.

The other area is the test cell itself, separated from the control area by a wall of three-sixteenths-inch steel plate and a steel door with windows halfway down, a door that looks as if it could lead to the kitchen of a suburban Colonial house but for the wire mesh embedded in the panes. Hanging waist high on a half dozen chain hoists from a steel beam in the test cell ceiling is a device that looks like part battering ram, part bomb, about the size of a large garbage can. It is what the 5K team has, so far, in the way of a Stirling engine.

One of Neill Lane's woes in life is that the pressure vessel of the 5K is about the size of only two familiar objects in the universe: a Hollywood movie bomb, such as would be dropped from an airplane, or, more prosaically, a garbage can. "It doesn't look *anything* like a *bomb*," he has said at least two thousand times. "And it doesn't look *anything* like a *garbage can*." To him, this is true, because he has spent the waking hours of half a decade designing Stirling engines. To the unenlightened, it looks something like a bomb. Or a garbage can.

The pressure vessel is domed at one end and flanged at the other; at the flanged end it is closed by a large steel cap and a ring of bolts. When the pressure vessel was fabricated, it was given a coat of paint so that it wouldn't rust or corrode; on the theory that light colors were slimming, the 5K team chose glossy white. Their efforts were for naught, because all they got for their trouble was a pressure vessel that Peggy Shank (who assigns most of the names and nicknames at Sunpower) immediately christened Moby. Moby's second name is left implied, although there have actually been references in scientific papers to the phallic imagery of Stirling engines, given that they are linear objects often encased in domed pressure vessels. The name simply adds another woe to Neill Lane's psyche: Sunpower has made engines called, acronymically, SPIKE and GRIZZY and SHARP, which are good names for prototypes; but it has also made prototypes given names like Fat Man, to which no engineer would like to lay claim. As a name, Moby Dick is marginally better than Fat Man.

Inside Moby at this moment are perhaps 50 percent of the parts necessary to make a Stirling engine and the alternator the engine is designed to drive; the pressure vessel is necessary because the engine is designed to operate using helium gas under a pressure about thirty-five times greater than atmospheric pressure, some 500 pounds per square inch. This also explains the three-sixteenths-inch steel plate on the wall and the wire mesh in the door windows, although if Moby fails along a weld at a pressure of 500-plus psi, no one expects to be saved by a little bit of sheet steel; instead, the pressure vessel was "overdesigned" by a large margin, and has already been tested to a pressure of nearly 900 psi.

What Neill and Abhijit and Dave Shade and several other members of the 5K team are testing on a cold and gray Athens, Ohio, afternoon represents nothing less than eight solid months of their lives. The first four months turned out to be a washout, and they started from scratch in September. So eight months of work was really compressed into four, which meant a lot of late nights, a lot of early mornings, a lot of Saturdays and Sundays. Many of those nights and Saturdays belonged to Neill Lane, who stands just outside the door of the test cell smoking a cigarette. He is not in the designated smoking area of Sunpower, Inc., but he can see it from where he stands, which counts for something. He stamps one foot, and then the other, against the cold. Neill is from South Africa, which, for all the bad one can say about it, has a climate infinitely superior to that of Athens, Ohio. "The people who settled this area must have come through on some rare spring day: the summers are horrendous; the winters are horrendous; spring and fall last about two weeks each," he says. "My feet are like ice."

Abhijit is also swaying from one foot to another as he confers with Dave Shade, going over one more time all that has been done so far, although Beej often sways back and forth, an unconscious habit that Shade and Hans Zwahlen, a technician on the 5K, never tire of imitating; Shade is forever threatening to calculate Abhijit's natural frequency; Hans complains that being around Beej makes him seasick. Abhijit is meticulous, fond of checklists and work plans, highly regarded for making lists of things that need to be done and then making sure that they *are* done. He checks the time frequently; he would like to call John Bean at Cummins and report a success, but he can't do this until they have one. Beej is responsible, above all, for serving as a liaison between CPG, which pays the bills, and Sunpower, which spends the money. It is Beej who calls John Bean at CPG regularly and keeps him apprised of Sunpower's progress, something he would dearly like to do this afternoon: eight months is a long time to pay the bills without knowing what one is getting.

Beej quizzes Dave Shade, the inflection in his voice rising up at the tail end of questions and ending, more often than not, with the "ay?" of Canada, where he grew up. "So we're ready to go, *ay?*" says Beej, to which Dave Shade says, "In a minute," as he stares at an oscilloscope. Neill Lane snuffs out his cigarette and comes back into the test cell, where he is joined by Joerg Seume, an engineer from Germany who is responsible for the design of a vital component of the engine they are about to test, and Barbara Fleck, the only native Ohioan on the team, who designed many of the test rig components themselves.

What they are about to do marks a unique moment in Sunpower's history in several ways. The machine that Dave Shade is ministering to, the machine designed by Neill and Joerg and Barbara as part of the project managed by Abhijit, is by definition unique, as all first prototypes are. The previous spring it was not exactly a clean sheet of paper, but it was a clean start on a project. Although it is still called, colloquially, the 5K, the project is now the "Five Kilowatt Solar II (Axisymmetric)."

The "II" in the name is a reminder that the original clean start was aborted. The original plan is now known simply as "the Box," because whereas a crucial component of the 5K II is "axisymmetric" (that is, round), the same component of the first 5K design was square, boxlike. The Box, when referred to at all, is referred to with a slight wince: five and a half engineers and a draftsperson spent four months on it; Cummins Power Generation, Inc., was billed several hundred thousand dollars for it, but it turned out to be, if not a dead end, at least such a tortuous path that it was finally abandoned. The Box is still not a happy topic of conversation around Sunpower.

So the 5K is unique as a first prototype; the test is unique because, although it makes perfect sense to do the test that is about to be done, and although such a test has been talked about as a good and rational way to proceed with a new engine design, Sunpower has never actually done such a test, a test of the resonance of the engine parts. That the 5K team is doing a resonance test gives a good insight into the management style of Neill and Abhijit; the test is something like the two-initial rule and the design reviews they insist on holding whenever they make a major change, so that anyone who knows anything can troubleshoot and play devil's advocate. All the most recent circumstances on the project conspired to whisper to the 5K team, *"Just* skip *the resonance test"*—circumstances like the false start that cost valuable time and lots of money, like a sponsor sitting next to a telephone waiting to be told that *this* design has been validated, like parts waiting in the machine shop to be fabricated, like a deadline looming. But Neill and Abhijit stood firm; they would have their resonance test.

Dave Shade is finally ready, having hooked up the test rig/engine to what seems like dozens of wires, cables, hoses, and umbilicals, and in the control area of the test cell turns a valve. The spectators don safety glasses (which would do nothing, of course, in the event of a total pressure vessel failure, but would keep dust and dirt out of one's eyes should the helium charging line come loose or spring a leak), and they all watch a pressure gauge as its needle begins to creep up. William Beale comes in, hands in his pockets, and peers into the test cell. "So what's happening?"

Neill answers. "We're just charging the machine for the resonance test." The pressure gauge measures in units of "bars," or atmospheres; one bar equals 14.7 pounds per square inch, which is atmospheric pressure at sea level; as Neill speaks, the needle passes 10 bar and creeps toward 20. For the resonance test, they will charge the machine to 35 bar and turn on the power. If they have done their work well—if they have done many thousands of calculations correctly, if they have translated those calculations correctly into the design of the parts, if the design was translated correctly into blueprints for the machine shop, if the machine shop has followed the drawings correctly and made the parts to specifications, if the parts were assembled according to plan, if the resonance test rig parts were also calculated, designed, drawn, and fabricated correctly—what will happen should be anticlimactic, completely and utterly predictable.

What they are hoping—expecting—to happen is this. The alternator inside the pressure vessel will be driven by electrical current from a transformer in the control room. The electrical current will cause one part of the alternator, a barrel-shaped ("axisymmetric") cylinder of magnets about twice the size of a three-pound can of coffee, to move back and forth, to oscillate in response to the alternating current flowing through it via the inner and outer iron rings of the alternator itself. Attached to the magnet cylinder is a flat piston, which consequently will also move back and forth. The piston is attached to a hollow rod that runs through the center of the cylinder of magnets and that of course also moves back and forth. At each end of the magnet/piston/rod assembly are large steel springs; because there is no displacer in the engine at this stage, the steel coil springs will provide some of the bounce that normally would be supplied by the displacer and its own gas spring.

Because the electricity to be delivered into the test rig is at a frequency of sixty cycles per second, the magnet assembly should bounce back and forth on the springs at sixty cycles per second; the more current fed into the machine, the longer the oscillations will be—if electrical current is thought of as analogous to pushing a child sitting in a playground swing, the higher the current, the harder the push, as the child in the swing moves back and forth. What the 5K team is looking for is resonance of the mechanical, magnetic, and electrical guts of their engine at sixty cycles per second of alternating current, alternating current high enough to push the piston assembly in and out at a stroke of about sixteen millimeters, or about half an inch. Eight millimeters out and eight millimeters back, sixty times a second.

Sometimes, children in swings will do tricks to make the swing motion jerky; they might tug on the chains, for example, or twist their bodies, or lift up in the seat at the extreme of the arc, so that the chains rattle and jerk. When they do such tricks, they disturb the normally smooth arc that swings seem to want to have; they disturb the resonance of their swinging. Engineers in virtually all fields other than that of Stirling engines design structures to *avoid* resonance. They build dissonance into the structure of buildings and bridges and towers so that they are *not* resonant, because wind, or the rotation of the earth, or any rhythmic motion at or near the resonant frequency of a structure can cause it to begin to resonate. And a resonant motion, once started in something the size of, say, the Brooklyn Bridge is hard to stop. Resonance is why soldiers marching or running in formation are taught to break step when crossing a bridge; resonance resistance is built into structures like athletic stadiums, so that when fifty thousand people get up at the same moment and start to march out, their steps don't start the stadium on its way toward shaking to ruin. Several years ago, two engineers calculated first the resonant frequency of the Empire State Building and then the frequency at which they would have to run back and forth on the observation deck in order to start the building swaying. Armed with only a stopwatch, they in fact managed to do so, although the building management took a dim view of something that most engineers think is pretty neat: two nerds with a calculator and a stopwatch making the Empire State Building sway like a willow. They stopped before they rocked it off the foundations. The resonant frequency of large and important objects must be calculated and accommodated; when it is not, catastrophe is inevitable. Most people have seen a film of the Tacoma Narrows Bridge, in Washington State, swaying and twisting like a Slinky before it collapsed from resonant motion set up by the wind; some of the early all-steel ships made in World War II using an entirely plausible (but unvetted for resonant frequency) new technology sank when something as seemingly inconsequential as a tool was dropped—the harmonics of the steel plates were such that the resonant vibration caused the ship to break apart.

But a Stirling engine needs to be resonant in order to run; more particularly, it needs to be resonant at the design frequency of sixty cycles per second, so that when it runs and produces electrical power (rather than absorbing it, as in the resonance test), it produces power at the frequency that electrical power is consumed. All structures are resonant at some frequency. The 5K needs to be resonant at a frequency of sixty cycles per second; it has been designed and computer simulated to be resonant at sixty cycles; the materials were chosen and fabricated with this in mind; all that is left now is to see whether, in fact, this is true. They hope to gain other information as well, information on the performance of the alternator, on the performance of the bearing surfaces on which the magnet assembly rides. But mostly, they hope to see whether the engine is resonant. They hope to both justify the past and divine the future, to see whether the change from the Box to the cylinder was indeed the right move, and to see whether the engine will indeed work. If the test rig is resonant, they will probably have a party; the party committee has been agitating for one for weeks, and while a successful resonance test doesn't exactly fit the criteria for having a party, it will probably be close enough.

If the test rig is not resonant, though, if it turns out they've done something wrong, they still have some time and money with which to try and recoup, although they are running low, this cold December afternoon, on goodwill. While no one in the room expects a *total* failure (Sunpower is justifiably famous for gleaning data from things that have just blown up), there are enough of what are occasionally called "downside probabilities" to make this test of much more than passing interest. An engine blowing up is one example of a downside probability; so is an engine or alternator that sticks, or burns up, or breaks into pieces, or shorts out. A far less catastrophic downside possibility is what might be thought of as another sort of test, not a test of the actual hardware but, rather, a test of the designers' abilities to perform a dynamic analysis, to translate that analysis into geometry, to translate the geometry into fabricatable drawings, to translate the drawings into parts: the 5K team is about to take its midterm examination.

Here's why. A free-piston Stirling engine is a device that must capitalize on the very features that make it unique in the first place—a piston and a displacer that move freely in space. Without any mechanical linkage to move the piston and the displacer through their cycles, the only things that will move them are the forces of the helium inside the pressure vessel. The piston and the displacer both (in the final machine) will "bounce" on what Sunpower refers to as "gas springs," terminology that conjures up the image of a coiled tube of gas, like a shape drawn with smoke. A gas spring, however, is simply the phenomenon that makes the handle of a bicycle pump "bounce" if one covers the outlet with a finger and then pushes on the handle: compress a gas, and it pushes, or springs back. In the 5K design, the volume of helium in front of the piston is concentrated on the piston face, making it a relatively small volume; behind the piston, or behind a piston without a gas spring, is a much larger volume and hence a different pressure, and a difference in pressure on either side of the piston will win out over an engineer's mathematical abilities every time, and the piston will tend, over time, to find a place where the pressures are equal.

The piston, in other words, seeks to find the place in the cylinder where the pressure of the gas behind it equals the pressure of the gas in front of it: the center of the gas pressure, rather than the center of the cylinder. In a cylinder where the pressure on one side of the piston is twice the pressure on the other, the piston will not stay in the physical center but migrate to a place where it will bounce back and forth between equal pressures. The piston will "center" in the place where the pressures are equal, no matter what.

So what the 5K team has done is make a piston gas spring. The piston gas spring doesn't look like anything, not even a coil of smoke. It doesn't look like anything at all, since it consists simply of the empty space behind the piston, a closed chamber that very closely approximates, in volume, the volume of the piston cylinder in front of the piston: equalize the volumes, and you center the piston.

Or so it would seem. The calculations for this are relatively straightforward (if complex), and Neill Lane has done them. In theory, a Stirling engine can be made to run perfectly, exactly, exquisitely in balance. In practice, though, come headaches. For example, in doing the calculations for a piston gas spring, one uses figures for the efficiency of the heat exchangers and the regenerator, figures that are real numbers out to three decimal places. But in practice the performance of the heat exchangers and the regenerator is determined by dozens of real-world factors impossible to account for—in the theoretical regenerator, the thousands of feet of steel foil wound in layers around the regenerator core are of an exact thickness and an exact distance apart. In the tangible regenerator, the steel foil was wound up by Abhijit and Hans and Paul Moore, by hand, using all manner of improvised jigs and tension wheels and crimping tools and dies and mandrels and spools with screwdrivers through them and ribbons of steel foil draped all over the lab—a process as exact as they could devise, but still one with a margin of error.

Likewise, the cold heat exchanger is as close to a piece of sculpture as one could make without wearing a beret, so handcrafted a piece of work it turns out to be: it consists of hundreds of copper fins made from copper tubing, working with which is a notoriously inexact science; soldering them together is another notoriously inexact science, and machining them into a solid copper ring is so inexact a process that many machinists are simply incapable of doing it. These real-world parts will not, except by coincidence, give theoretical numbers, but it is with theoretical numbers that empty spaces like the piston gas spring must be designed.

Thus, the resonance test, which is also in part a centering test: where will the gas spring on the piston cause the piston to find its center? If the piston runs too far "out" (or away from the heater head), the engine will lack power, because "power" is determined in large part by the force of the hot helium on the piston face; if the piston runs "in," or toward the heater head, the engine will also lack power, unable to run at full stroke, because the piston would then bang into the displacer. Given that the piston travels only about fifteen millimeters out and fifteen millimeters in (a total distance of about an inch and a quarter) when the engine is at full stroke, being a couple of millimeters off center can easily make Moby Dick become Moby Pig: underpowered in one direction or the other. In past programs at Sunpower, centering the piston has always followed an inevitable route: design an engine that centers theoretically, run it, and then add some method to center the piston. "Centerports," for example, are grooves on the piston that move past holes in the piston cylinder as the piston follows its tendency to find its center—in other words, intentional leaks between the "work space" in front of the piston, and the "bounce space" behind it. But centerports use up power, because they in essence vent helium from the place where it is capable of doing work into a place where helium absorbs work: a double loss. Matching volumes in the work space and the bounce space promises, in theory, at least, less of a potential loss.

The needle on the pressure gauge approaches 35 bar, and Dave Shade turns the helium off and the electrical power on. Current will be fed into the test rig at low levels first, then progressively higher, through a device called a Variac—a *variable AC* transformer. Beej throws some switches, checks his clipboard, and looks at Neill. Neill looks at Dave Shade, who shrugs. "Are we ready?" asks Neill. "*We* are," says Dave

Shade, with a grin; he's not going to presume anything about the test rig, having learned from more than a decade of experience with Stirling engines that the bad-luck demons just *wait* for someone to get smug.

So Neill looks at Abhijit, and Abhijit looks at Neill. "*You guys* put it together," Neill says with a laugh. "Let's do it, *ay?*" says Beej, with a grin. And Dave Shade turns the current on.

On the oscilloscope screen, a flat green line appears, representing the amplitude of the piston inside the pressure vessel; as Shade sends more current through the alternator, the line grows, approaching ten millimeters—as far as they had planned to go in the resonance test. William comes in, hands in his pockets, peers at the oscilloscope for a moment, shrugs his shoulders, and goes back out; Neill and Beej and Dave Shade continue to stare at the scope, although the line is emphatically uninteresting. The engine—or rather, the parts fabricated and assembled thus far—are doing just what they are supposed to do, which is nothing more than move back and forth. Put this way, it seems trivial; at some future time, when alternators and Stirling engines are bought and sold as commodity items off an assembly line, it will be trivial. But at this moment it represents a great leap forward for the 5K people: they aren't making electricity (they are, in fact, consuming it), but they have something that will make electricity, and it seems to work. This is a major part of what the resonance test is supposed to test. On the screen of the scope, the green line is not centered exactly: the piston is, in fact, running in slightly, but since this is not only a new engine but a new test, no one is willing to say that the piston will be exactly centered in the test, or that the piston will run when the engine runs, exactly where the piston ran in the resonance test. But as a validation of the design thus far, the test is a success: as Neill and Barbara and Joerg and Abhijit and Dave Shade stand and watch the green line, watch the numbers change on the data acquisition screen, they can be forgiven for being more than slightly pleased: Beej reads numbers off the computer screen and does some calculations; Robert Redlich, who did the electrical design of the alternator, does calculations as well and smiles broadly. "If Robert's happy, then I'm happy," Neill says. Although everyone's attention is focused on the oscilloscope screen and the data acquisition computer screen, the real star of the test is encased in the pressure vessel: the linear alternator.

A mechanical device, like a Stirling engine, if it is said to "do work," has to *do* something: there has to be a way for the power produced by the engine to enter the real world, to lift a weight, to pump water, to turn a shaft, to make electricity. An engine has to be cyclic, and it has to do work. William Beale's first free-piston engines were water pumps, because they were in essence reciprocating structures that moved up and down, which is how reciprocating water pumps work, and which is what caused William to think of the Third World, rather than the First or Second, because developing countries are places where water pumping is an important, life-sustaining activity that has no infrastructure built to do it. But a water pump is only so interesting in, say, the United States, where, as everyone knows, water simply pours out of faucets.

Likewise, the United States already has a plethora of devices for lifting weights and turning shafts; these are processes that have the benefit of an infrastructure already set up handsomely to use rotational, rather than reciprocating work. Electricity, too, has an immense infrastructure, but with one crucial difference: electricity is finally that which moves through a wire, and electricity produced by a reciprocating alternator is indistinguishable from electricity produced by a rotating one, just as electricity produced by a nuclear reactor is indistinguishable from electricity produced by a coal-fired power plant. So the linear alternator was born.

A linear alternator works much the way a conventional, rotating alternator does, which is to say in a fashion generally considered to be opaque, although in its simplest form an alternator is simplicity itself. Electricity comes from the interplay of magnets, copper wire windings, and iron. A magnet "induces" a magnetic field when it moves in the space between two pieces of iron; when the magnet moves in one direction, the polarity of the field runs in that direction; when the magnet moves back, the polarity of the magnetic field changes as well. (This is the "alternating" of "alternating current," or AC.) The copper wire, an excellent conductor of electricity, serves to carry the current induced by the movement of the magnet through the gap between the two pieces of iron. The science of magnetism and electricity can be absorbing, and electrical engineers work in a polarized world of positive and negative; most mechanical engineers are content to say, as Neill Lane does, that "the movement of the magnets 'causes' electricity to flow; see Faraday, 1831." (Michael Faraday, though less well known now than scientists like Watt and Newton, is the father of electricity as the contemporary world knows it.)

Everyone knows that magnets have a "positive" pole and a "negative" pole, if only because everyone has seen a magnetic compass, which points helpfully toward the North Pole. When a magnet is passed between two pieces of iron, the magnet induces a voltage (which can be thought of as the force that pushes electricity through a conductor such as copper wire). Moving the magnet in one direction induces a voltage with a certain polarity, either positive or negative; moving the magnet in the other direction induces a voltage of the opposite polarity. This is electromagnetic induction; when it occurs in a closed circuit, a current flows, and a voltage causing a current to flow is what we call electricity when we plug in a lamp.

Current is generally produced by the round-and-round spinning of either magnets on a rotor or windings of copper on a rotor inside a fixed ring of magnets. (The alternator on an automobile, which charges the battery, is of a typical alternator design.) A linear alternator is constructed in such a way that the current is induced by moving the magnets back and forth in the gap between two cylinders of iron, rather spinning the magnets around and around. In a linear alternator, the magnets are the moving parts, because compared with the mass of the iron cylinders between which they pass, they are comparatively light and thus take less energy to move. (Think of three nested cylinders, the largest of iron, the next, magnets, the innermost, more iron.) The magnets in the 5K alternator are arranged in three rows end to end, and assembled into rings of 56 magnets: 168 magnets in all, each about a half inch thick, a half inch tall, and two inches long. At full engine stroke, each row of magnets induces a voltage in the iron cylinders; this, in essence, triples the potential power of the alternator.

The inner and outer iron cylinders, rather than being solid, are in fact made up of many thin laminations, held in a cylindrical shape by glue; the inner laminations, of which there are nearly a thousand, fit closely inside the cylinder of magnets; the outer laminations, of which there are more than two thousand, fit closely outside. If the inner and outer iron were simply solid cylinders, not much electricity would be produced, because of several effects peculiar to magnetism and electricity: magnetic "eddy currents" in the iron, "saturation" of the iron with magnetic lines of force, and so on. The electrical potential of the alternator has been calculated and evaluated with all of this in mind, but the principle, the theory, is the same as that demonstrated by Faraday. The magnets move between the laminations and cause electricity to flow. The final step is to have many loops of copper wire in "windows," or grooves, of the outer laminations; copper is a good conductor of electricity, and the copper windings transmit the power produced by the alternator; each turn of copper wire carries a voltage, and the voltages are cumulative. But it's all finally just magnets, iron, and copper.

From this, electricity is "caused," and the world as we know it hums along.

Put this way, making an alternator, even a linear alternator, sounds trivial. But all of the requirements, the constraints, the criteria of the design, conspire to make it anything but trivial. Conventional alternator companies looked at the design specifications and said, "No way." Magnet companies said, "Nope." Transformer factories said, "Sorry." What scared them all off (and what forced Sunpower people to do much of the fabrication of the alternator themselves) was simple: companies that make alternators and magnets and transformers don't make anything reciprocating. Hardly anyone does, because there is little call for it. There is little call for it, because it is so technically demanding. This is a perfect example of the vicious circle that R & D companies face often and that Sunpower faces all the time: if General Electric or General Motors doesn't already use it, it's likely that it doesn't exist. If it doesn't exist, someone has to make it, and if someone has to make it, it's going to be expensive if not impossible, or at least unheard of. Things that we take completely for granted nowadays were once impossibly clumsy and jury-rigged prototypes: the first VCRs took two people to carry; the first cellular phones required a trunkful of electronic equipment. Likewise with the fabrication of an alternator that doesn't resemble any alternators that the big boys already buy.

The embodiment of the challenge Sunpower faces is the magnet assembly—the magnet "paddle"—in the test rig. Magnet manufacture is an art *and* a science, because magnets are made in a sort of witches' brew series of steps that often result in variation in the strength of two apparently identical magnets; magnets are susceptible to temperature and electricity and rough handling, and the more powerful the magnets, the more susceptible, the more delicate, they can be. The magnets that Sunpower uses are very expensive and very strong; scrap magnets, broken magnets (they are very brittle), and leftover magnets from other projects are stuck to just about every metal surface at Sunpower, and are nigh impossible to simply pluck up—Dave Shade calls them the most powerful refrigerator magnets in the world. Made of an alloy of iron and neodymium, they are powerful enough to stop a watch, erase a computer disk, freeze a calculator.

There are several reasons that there are 168 magnets in the 5K instead of 3 or 4 very large ones. One is simply polarity; a curved magnet, for example, cannot have its positive and negative poles in the same places as a flat one can; the curvature affects the polarity. So, many small, flat magnets are arranged in a cylindrical shape.

Another reason for so many magnets has to do with a design decision made to squeeze more power out of an alternator of a certain physical size: the 5K's alternator is "triple-windowed," or actually three alternators in one. Each of the three rings of magnets passes between its own set of inner and outer laminations, which is complicated to do electrically and geometrically; it means the magnets must have an orientation both to the laminations (as in any alternator) and to one another. The magnets must be arranged with their positive poles facing one set of laminations, and their negative poles facing the other, and then glued together, glue being the only nonferrous and gentle fastener that has a hope of holding magnets together. But since like poles repel and opposites attract, each magnet always wants to jump away from the magnet it should be next to and attach itself instead to its neighbor's underside.

When Hans Zwahlen, one of the 5K's technicians, started laying out the magnets for the alternator, he was like a man chasing mice: put one next to another, and the first one leaps away. And hard as it is to arrange the magnets on a workbench, then in a canister-like jig so that they can be glued together, the magnet paddle, with its 168 magnets, will ultimately be moving at sixty cycles per second, fifteen millimeters in 1/120 of a second in one direction, then fifteen millimeters in 1/120 of a second in the other direction—about three meters per second—when the engine is running, so the forces of inertia and acceleration must also be considered, and they are huge. To put it the way it was put to magnet and alternator manufacturers, Sunpower needed a magnet paddle with a large number of very strong magnets, held together in a frame that was light (so the magnet paddle wouldn't take a lot of the engine's energy just to move it) and nonferrous (so it didn't interfere with electricity generation), that could be put into a pressure vessel and oscillated at sixty cycles per second for several thousand hours without breaking apart under very strong forces. No one wanted to build such a thing, so Sunpower had to.

But first there was a very expensive false start, and a series of feints and program changes that fill two huge binders of correspondence on Beej's bookshelf. Even with such exhaustive documentation, there are several versions of the history, some frankly revisionist, some sounding like conspiracy theory. At one point in the back-and-forth and round-and-round discussions on the fundamental design, at least one engineer on the 5K, tugged in different directions to the point of distraction, didn't get right to work on what was supposed to be the definite plan, because recent experience suggested that the definite plan would be changed. So he did nothing for several days while looking industrious and put upon, and sure enough, he was right, because the definite plan was changed, and he was given a new brief to get right to work on. He would have had to throw out several days of work, if he had done it. Even in loose organizations like Sunpower, this sort of thing is a bad sign, although it does add to the Sunpower legend. Compressed and objectified (which means that no one will see the story quite this way), the startup on the 5K was something like this.

The original alternator design for the 5K—the *original* original (it would have to be called retrospectively something like the "Solar Five Kilowatt Minus III")—was an axisymmetric design for a nine-kilowatt machine. This was bigger than Sunpower had ever made before, but Neill Lane and Jarlath McEntee had just finished the twenty-five-kilowatt NASA design, which was an axisymmetric scheme. (The NASA engine was never built.) Sunpower engineers had gotten pretty good at making axisymmetric alternators, although they had never made one like the one they would need to make for the 5K—or the 9K, as it were; the most efficient alternator was on the SHARP, which was nearer to two kilowatts than to nine, but the design principles had been validated. Neill Lane did a preliminary design for the 5K—the 9K—and then took a long-planned trip back to South Africa. While he was gone, much back-and-forth with Cummins led to a series of discussions and decisions and negotiations about the ultimate goals and feasibilities of the project. For a while, the mutual understanding was that Sunpower would do the big engine, the 9K; for a while, it was that Sunpower would do a five-kilowatt engine as a precursor to a bigger machine, an intermediate step. Much of the negotiations had to do with money. Would Cummins get the 5K and the 9K for one price—its original million dollars? Or would it get one or the other for the million, and need to pony up another million for the other engine? The idea of two engines for one million was planted—*somehow*—in CPG's thinking during this process, as was the idea of two projects for two million in Sunpower's collective consciousness. Both had to be disabused of what were, for obvious reasons, attractive scenarios. For a few weeks (there was always someone who knew exactly how things were to go, or who needed to sign off on the final arrangements, who also had to take a trip and thus hold things up for a few days), the size of the engine changed like the stock market, up one afternoon, down the next morning. And then the Box crept in.

The Box alternator idea was the result of a suggestion to Cummins by William, a suggestion that itself was the result of William's never-ending quest to make Stirling engines *(a)* out of readily available components and *(b)* seem less exotic and more familiar to the real world. The logic was that the Box alternator looked, cursorily, like something that exists in the real world, an electrical component called a transformer. The idea was to use relatively large, flat magnets—each magnet in the Box design was about the size of a pack of baseball cards—arranged in two plates, each about the size of an LP record; the laminations for the alternator would also be flat, arrayed in three layers, one on top, one on the bottom, and one between the two layers of magnets. This notion was based on the long view, toward the days when Cummins would like to manufacture a thousand, or many thousands of, engines; William Beale always looks for devices in the real world that can be adapted, or that at least appear familiar, for Stirling engines, in an effort to overcome the no-way-can-we-make-*that* response of manufacturers who have never seen a Stirling engine. The Box idea was predicated on appearances, mostly; transformers are common and well understood, but they have no moving parts, and they have no magnets. Transformers are devices that change the voltage of electrical power and, as such, have a place in the electric utility industry, the power conversion industry, the model-train industry (it's that box you plug in the wall when you run the train around the Christmas tree). Transformers do have laminations and copper windings and are available, or can be made, in virtually any size or configuration imaginable.

Neill Lane returned from South Africa to an engine design that had undergone a transformation so fundamental that it is fair to say that it was unrecognizable. He lobbied William; he lobbied Abhijit. He got nowhere. The arguments in favor of the Box were, first, that the sponsor had approved it, second, that a lot of the electrical calculations for it had already been done, and third, that it would work, thus ultimately saving Sunpower and Cummins much time and money on down the road. Just call up Acme Transformer and order some alternators.

So the 5K team buckled down and tried. The first tactical problem that arose was one that dogged the entire design process: the Box was impossible to visualize, to draw, to "see" inside; it was, inadvertently, a magic box.

A necessary part of the design process is being able to visualize a three-dimensional structure from looking at a drawing done in two dimensions. Symmetrical objects are those that have two identical "halves" when divided through any plane: a soup can has an "axis of symmetry" around which everything is mirrored: the top is a mirror of the bottom and the two halves mirrors of each other. Two drawings are necessary to illustrate the shape of a soup can, but only two: a side view and an end view. An object like a shoe box is not symmetrical in this way; it requires three views to describe fully its shape: end, side, and top.

Such a distinction seems trivial; good engineers and good drafts-persons know how to overcome this, of course (otherwise, we would have only symmetrical objects around us: no violins, no bicycle frames, no trapezoidal cans of corned beef), but the descriptive geometry involved is not trivial at all: it is time-consuming, highly iterative, imaginatively taxing, and fraught with potential errors. In order to calculate and draw the geometry of the Box alternator, the 5K team members spent many, many hours figuratively (and occasionally, literally) craning their necks, trying to "see" the view of the Box that they needed to draw. Drawing the Box turned out to be a lesser problem, anyway. Neill and Todd Cale, the team's computer draftsman, came up with a set of drawings that looked, to the unschooled eye, all right, including one top view that was good enough for presentations in meetings, although it was a sort of Potemkin village of a drawing; there was nothing "inside." Ironically, it was this drawing of a fantasy that ended up in Cummins's publicity about the solar project, thousands of flashy colorful brochures containing a drawing of something that Sunpower would later desperately try to forget. But it wasn't the drawing that was the showstopper. The showstopper turned out to be the large, flat sheets of magnets, which were supposed to be the easy part. They turned out to be impossible.

For an alternator to work, the magnets need to pass through a narrow gap between the laminations; the narrower, and more evenly spaced, the better, but at the very least, a fairly narrow and fairly even gap, the "air gap." If the air gap is too large, the electromagnetic induction is weak or nonexistent, because the inductance decreases by the square of the increase in the air gap: triple the air gap, and decrease the inductance by a factor of nine. No gap at all would be worse; if the magnets rub on the laminations, they lose much of their strength, and they lose it very rapidly; iron-neodymium magnets are demagnetized above temperatures of about 150 degrees Fahrenheit, and rubbing two pieces of metal against one another sixty times per second will generate a temperature like this in minutes.

So the trick was to design a large and flat plate of magnets that would hold together under the forces of acceleration and inertia in the engine, all the while the plate was trying to fold itself up along its glue joints. If it seems ridiculous to be using glue to hold together a crucial part in a million-dollar project, it is because the strength and the fragility of the magnets Sunpower needs to use are hard to overemphasize. The magnets can't be bolted or screwed together, because they can't have holes drilled in them: they shatter. They're hard to encase in any sort of frame, because a frame strong enough to hold the magnets flat needs to be too big to ride through the air gap. Magnets can't be attached together by means of any conventional metal-joining technique; they can't be welded, because the heat would demagnetize the magnets; they can't be brazed or soldered or riveted or taped together with duct tape. That leaves glue.

The attempts at a design solution for the flat magnet paddle, Neill Lane says matter-of-factly, would never have worked; he should know, of course, since he spent months designing the attempts. The plan was to build the alternator frame out of long, thin titanium rails and a hard polycarbonate called G-10 (which resembles superficially the hard plastic that telephones used to be made from): the titanium for strength, the G-10 for support, both because they are nonferrous. The laminations and windings would be supported by this. The weak link in this design was still, however, the glue in the magnet joints, because the frame that would attach the magnets to the piston assembly could hold only the plate of magnets in place; it couldn't possibly be strong and stiff enough to hold the magnets together.

Abhijit worked for weeks with the magnet manufacturer on the design of the magnet plates. The people at the magnet company were reluctant to have anything to do with the magnet plates; they kept offering suggestions that would make the plates easier to produce, but that would also keep them from fitting in the alternator. They didn't like the glue, they didn't like the tolerances, they didn't like the way the magnets had a tendency to fold up like Venetian blinds before the glue dried. Finally, with much ceremony and fanfare, they shipped the first magnet plate via special carrier to Sunpower, and the shipping crate containing it was carried, with equal ceremony, to a specially prepared area in the laboratory, an area notable for a large plywood booth with a hinged door, festooned with signs warning of strong magnets and delicate assemblies within, a booth built specifically for this plate of magnets. Abhijit opened the packing crate containing the single intact plate that the magnet company had thus far (in several months) been able to produce. It was still intact.

Beej wiggled his fingers like a magician and gently lifted the plate out of the container. So far, so good. The first thing he wanted to check was the dimensions of the plate, which was roughly the size of an LP record album jacket, although about twice as thick; how close was it to the specified size? As he was considering the best way to proceed, one of the test engineers came over to help with what looked like a delicate but fairly straightforward operation, an operation that also looked as if it could use an extra pair of hands. Beej was holding the magnet plate in front of him, contemplating, considering, planning his approach, when the test engineer said, "Here, let me give you a hand," picked up a large pair of vernier calipers, and brought them toward the magnet plate.

Vernier calipers are lovely precision instruments, tooled carefully out of hard steel that holds its size and shape well, is resistant to gouges and nicks, and is machinable to very fine tolerances. A good set of vernier calipers can last fifty years if taken care of properly. A set made out of, say, aluminum would rapidly become an inaccurate and imprecise device, one that no good engineer or machinist would want to touch. Good vernier calipers, on the other hand, are a delight to handle; they have heft and balance and a nice finish. Good steel makes good tools.

Good steel is also magnetic. What happened next takes far longer to describe than it took to occur. Beej saw it happening before him; his shout of warning never made it out of his throat. The test engineer moved the calipers toward the magnet plate until the calipers entered the magnetic field. Then the magnets moved the calipers and the test engineer. The jaws of the vernier calipers cracked the magnet paddle into a dozen pieces along its glue joints, neatly, quickly, and completely. The magnet plate, which was intended to last through a series of tests and engine runs totaling something well over one hundred hours, had survived outside its packing crate for only a matter of minutes. Most of the members of the 5K team hadn't even gotten to see it.

The test engineer (it was Al Schubert, a gruff and iconoclastic soul who believes religiously in good, stout, solid design) was apologetic, but also philosophical; anything that fragile was not a good idea, particularly for such an integral part. Everyone agreed that the magnet plates would never have worked; if they didn't break when they were taken out of the box, they would break soon thereafter, and better to break sooner rather than later, when the magnet assembly was in the engine or, worse, when the engine was running and the broken pieces would be in a perfect position to destroy the alternator and perhaps the engine itself.

But simply having a crucial component break into pieces without damaging anything else could hardly be considered success, or even progress. Cummins had already invested more than a quarter of a million dollars—about 25 percent of the money budgeted for the entire program, so Abhijit couldn't simply call them up and say, "No can do." The many other parts of the engine and alternator couldn't be made, or even designed, until there was a solution to the Box dilemma, but the only plausible solution to that dilemma was the very one that no one wanted to hear about: abandonment.

But the facts, as Neill and, eventually, Abhijit saw them, were plain: they had spent four months on the Box and had plenty of headaches to show for it, but little else. When they tried to make little lists of pros and cons for the Box, they ended up with very lopsided tallies. (One of the "pros," if it could be called that, was that many thousands of dollars of titanium and G-10 had been delivered with which to make the alternator, a huge stack of expensive material that the 5K team members averted their eyes from whenever they walked through the machine shop.) The only alternative to the Box that anyone could come up with was an axisymmetric design, but for an alternator the size of the one they would need to build, they would be starting from scratch.

A full-scale meeting, with William Beale and the engineers from Cummins, was scheduled, in which Neill and Abhijit would make a presentation and, they hoped, be able to make a good, albeit painful, case for abandoning the Box; Cummins would have to be told to forget about the Box, and thus they would all have to move forward at a much accelerated pace: the deadline for a working engine was absolutely firm, because Cummins already had contracts to sell the three engines that would follow the first prototype, and promised delivery; a whole plan, involving much time, many people, and large amounts of money, was already in place.

It took a fair amount of courage to arrange such a meeting, let alone be the bearers of the bad news. But the 5K team saw no other solution.

The meeting was scheduled at Sunpower for a Wednesday afternoon in August 1990, a hot and sunny day. As they were wont to do on sunny days (even on days with big meetings), the younger Sunpower employees, Neill and Eric Bakeman and Jarlath and a half dozen others, drove up to the park at lunchtime to play a silly and exhilarating game called ultimate Frisbee, which is something like a cross between basketball and soccer, with hints of rugby, only played with a Frisbee instead of a ball.

And Neill Lane, who plays ultimate with great intensity, dove for an errant toss and, in doing so, shattered his right forearm just above the wrist. Instead of spending the afternoon in a meeting with Cummins, he spent it in the emergency room at O'Bleness Hospital, in Athens, being given what he wistfully referred to later as "very good drugs" and having his wrist pinned. On Friday, while he was still at home in bed, John Crawford, Sunpower's senior engineering manager, came to see him. "The Box is dead," John told him. "We need a new design."

Chapter 7

Back to the Drawing Board

The sponsor is *not* the enemy; the sponsor is *not* the enemy.
—*Abhijit Karandikar, in the heat of a design review*

Then why does it always seem we're fighting a losing battle?
—*Neill Lane's response*

THE EARLY DRAWINGS for what was to be the "5 Kilowatt Solar II" engine—the axisymmetric design—are identifiable by their dates and part numbers, of course, but also by the spidery and odd scrawl of Neill Lane's initials, the sort of scrawl one might expect would be made by someone wearing a cumbersome cast on his wrist and forearm. Neill Lane spent the early weeks of the new design effort with his arm in such a cast, and his initials on the drawings betray it. But there was not time to wait.

In retrospect, it seems impossible to imagine any design but an axisymmetric one for the Cummins engine: the project required a leap of several orders of magnitude from anything that Sunpower had ever envisioned and built in terms of power density and efficiency, and it had to be conceived of as an engine that could be built in large numbers by a company other than Sunpower—Cummins, after all, was an engine manufacturer above all else, and its people were extremely wary of ending up with a design that could be executed only by Sunpower (which would make the project a very expensive gadget or sculpture, but not, in Cummins's terms, a real piece of engineering) or that couldn't be executed by anyone (which would make the project a disaster).

William's opinion of the perfect sponsor for a development project was well known: the sponsor puts up large amounts of money, waits patiently, and in the end is rewarded with a prototype that is Sunpower's interpretation of what the sponsor should want—William has said more than once that he couldn't understand why the sponsor would not simply leave Sunpower alone to do what it did best, and not interfere, hence his statement, moments after the very founding of the company, that he wanted Sunpower to be working on "things that people need even if they don't know they need it yet." This was not what Cummins had in mind or what Neill and the 5K team wanted to do. To get beyond the Box, the team had to start, if not from scratch, at least from a breathtaking disadvantage in terms of actually building a production machine. Neill's broken wrist served as a memorable marker for the 5K calendar: it was, for all practical purposes, day one all over again.

Engine design—at least according to the Witwatersrand-trained design engineers like Neill and David Berchowitz—moves from two concrete extremes through a thicket of calculations, simulations, sketches, and drawings to meet in the middle, where an engine should blossom, the inevitable result of thermodynamic rules.

The extremes are, at one end, Power In, and, at the other, Power Out. Power Out is what the sponsor buys, a machine that produces x kilowatts of electrical power, in this case nominally five kilowatts. Power In is what an average sunny day in a place like the West Texas desert (site of the Cummins solar dish test facility) produces when the rays of the sun are captured on a solar concentrator made up of twenty-four mirrorlike dishes and focused on the heater head of the engine: in the case of the solar concentrator Cummins will use, Power In is thirty-four kilowatts. Catch the sun's rays on mirrors, focus the rays on the heater head of the engine, get five kilowatts out of a pair of wires. A fairly energy-efficient American house uses around seven hundred watts of electricity, steady state, around the clock, so seven houses would have "free" electricity, less of course the amortization of the cost of the dish and the engine. QED. The rest is in the details.

But what details they are! What sort of design will last forty thousand operating hours? (Since the engine, obviously, will run only during daylight, this is about thirteen years.) What sort of design will last four thousand hours "mean time between failures"?—in other words, break down only every fifteen months or so? Given that every material that an engineer can specify for every part of the engine has a cost, what is the balance between cost and durability? Durability and weight?

Weight and stiffness? Stiffness and strength? Cost and strength? Cost and stiffness? Stiffness, durability, and cost? If the piston is made out of, say, titanium (a very expensive metal), does the increased strength (which means the part can be thinner) offset the decreased stiffness (which means the part must be stouter) when considered as a function of cost (versus, say, that of aluminum)?

Consider fabrication: Should this part be welded to that one? Brazed? Bolted? Sealed with an O-ring gasket or sealed with a weld? If part A is bolted to part B, can part C be attached without part A getting in the way of the wrench needed to tighten the bolts? Should part A be glued to part B? At what temperature does the glue melt? How long will the glue hold? If part A and part B and part C are all supposed to be aligned one with the other within a tolerance of, say, one-thousandth of an inch, are the tolerances between them cumulative or individual? If the tolerance is one-thousandth of an inch, and the material grows in size when heated, what is the actual tolerance at operating temperature, and should a material be chosen that expands proportionately less when heated, or should it be chosen for stiffness, or strength, or cost, and the other parts chosen and designed around it? "God," says Neill Lane, "is not in the details. The designer is in the details."

When Neill starts on the axisymmetric design, he starts with several knowns and several broad parameters. He also has the benefit of his own and his colleagues' vast practical experience with Stirling machines. From this experience, for example, he knows that at least three showstoppers are possible in any design, three pressure points that will corrupt the most methodical design process.

One is the hot end of the engine, the heater head, the part of the engine that will be heated to 675 degrees Celsius (about 1,400 degrees Fahrenheit) while containing helium gas at a pressure of about 40 atmospheres, or 600 pounds per square inch. This is as hard as it sounds in and of itself, but because of the necessities of heat transfer, the heater head must include several hundred small steel tubes through which the helium can flow: it can't just be a block of metal; further, the heater head must accommodate one of the most difficult environmental conditions a materials scientist could envision: it must heat up to 675 degrees Celsius in the morning, stay that hot for eight hours, then cool down to ambient temperatures overnight, and then heat back up again. The cumulative stress of such a heating and cooling cycle is tremendous.

For a heater head to do this thousands of times is extremely difficult to design for, in part because of the peculiarities of metals and how they react to temperature and the forces of the gas.

For example, a layperson might think to simply make the heater head really *thick:* thickness equals strength, right? Perversely, thickness is the *worst* enemy in a vessel under extremes of pressure and temperature: the more material, the more material there is to expand because of heat, and the forces of this expansion are cumulative, and not even cumulative linearly. In certain situations, the forces are the square of the increase; in others, they are the cube. So making a heater head twice as thick might increase the internal forces by as much as a factor of eight. In this direction lies certain failure. On the other hand, the heater head can't be too thin, or it won't be strong enough to contain the 600 psi of helium. And even when designed through painstaking analysis and calculation, the results are often unpredictable, in large part because a Stirling-engine heater head is not the sort of object that high-strength metal alloys are invented for—alloy manufacturers seldom know what a Stirling-engine heater head might even be.

So Sunpower must always extrapolate data on materials that was collected with other purposes in mind, and this can be disappointing. One heater head design for the 5K, which took Neill and Jarlath McEntee several weeks of computer analysis to come up with and several weeks of fabrication to make, lasted for about ten hours under operating conditions before it began to change shape as a result of the tremendous forces it was subjected to—Jarlath later calculated that the pent-up forces in the heater head from just those few hours of operation were something on the order of 800 megapascals, a truly astonishing number, given that one pascal is one "newton" of force per square meter, approximately equivalent (in the mnemonic device of the engineering student) to the mass of one apple sitting on a square yard of carpet. In ten hours, the combination of high temperature and high pressure had thus generated an internal stress within the head along the lines of 800 million apples concentrated on an area about the size of the floor space in a phone booth.

Another showstopper in a Stirling machine that must be addressed from the very beginning has to do with the contact surfaces, any place where a moving part touches a stationary part: the piston in its cylinder, the displacer in its cylinder, the round puck of the displacer's gas spring in *its* cylinder, and, preeminently, the bearing surfaces that support the magnet paddle as it moves back and forth sixty times a second. One consequence of a moving part rubbing against a stationary part is the heat generated by the rubbing, which can cause the materials to expand and distort; another consequence is the wear that the softer of the two surfaces suffers; the third consequence is the friction inherent in rubbing two pieces of material together: much of the engine's power would be sapped just overcoming such friction.

The oldest and easiest solution to friction and wear on contact surfaces is obvious: oil or grease, judiciously applied, will permit a lifetime of moving back and forth. The world slides along on petroleum-based lubricants. Unfortunately, Stirling engines can't have oil or grease in them of any kind, at all, for a reason that should be obvious to anyone who knows anything about diesel engines. Any petroleum-based substance, at some convergence of temperature and pressure, will explode and burn. Diesel engines have no spark plug; instead, the piston in a diesel compresses the vaporous fuel until it explodes. So a bit of grease on the load-bearing surfaces in a Stirling engine would also explode and burn—so, no grease. Another problem in contact-surface design is highly Stirling specific: the motion in a Stirling engine is reciprocal, rather than rotational, and while in the real world there are a fair number of lubrication-free bearing surfaces for rotating devices, there are none for reciprocating ones. The mechanical and dynamic differences between rotating and reciprocating motions are profound, as the 5K team would time and time again discover.

The third issue that can stop a Stirling engine in its tracks is more ethereal, but still real. It is a catchall category called "losses," several dozen factors that are the result of the unusual dynamics inside a Stirling engine. Some, such as the amount of power it will take for the engine to overcome the friction in its bearings (whatever its bearing surfaces eventually turn out to be), are relatively predictable. Others, such as the amount of potential power an engine will lose as a result of the heat generated by helium being moved back and forth through small spaces, are calculable to a high degree of accuracy if the engineer has a comprehensive mental image of the path of the helium through the engine and of all the surfaces it touches. Many others such losses are small, but cumulative—the amount of energy dissipated when the helium passes through the regenerator, the losses caused by conduction of heat through the various cylinder walls, and so on.

Still other losses are predictably small in theory, but turn out to be large in practice as a result of imperfections in machining or fabrication. And still others are simply surprises. The designers predict, or try to, but some things they won't know about until the engine is built and running, if it runs at all—losses can be tracked down and addressed, but the engine must be running in order to do so, and some losses are so profound that they might keep the engine from its maiden run for days, or weeks, or months. Or forever. Some of the more expensive and elaborate Stirling programs in history had this result, which contributed to the Stirling's doubtful reputation as a potential power source, and to the widely held opinion the a Stirling engine is a bitch to design.

Engineers of the "theory of design" school, like Neill Lane (as opposed to the far more common "calculate, then fabricate" school), believe in approaching issues such as these methodically and thoroughly, with an eye toward science as much as toward commercialization: part of the original 5K plan with Cummins included money and time for a complete test program for the bearing surfaces, upon which the piston's mass would be suspended, and for the sealing surfaces, which would be used around the piston and displacer to make for tight fits in their respective cylinders.

Joerg Seume thus planned a bearings and seals test rig to evaluate all manner of materials and designs for the program, a task for which he was well suited. Joerg is well trained (in Germany and the United States, where he received a doctorate in mechanical engineering from the University of Minnesota), meticulous, and, above all, methodical. He came up with a three-step plan for the 5K bearing surfaces, the first of which was a set of bearings to be used only in early tests, with the second and third stages requiring the design and fabrication of relatively exotic and rather expensive materials that had done well in rotational uses but had never been tried in reciprocating machines. He was in uncharted territory, which he liked, and he was "doing science," which he also liked, and which at Sunpower was often cast to the wayside under the pressures of deadlines and the shortfalls of funding. Joerg and two graduate students from Ohio University, funded in part by a grant from the Edison Materials Technology Center (EMTEC), in Dayton, Ohio, and operating under the auspices of William Beale's last, tenuous connection to OU, the Center for Stirling Technology Research (CSTR, pronounced "sister"), moved into a test cell, began doing calculations and drawings, and set up a timetable for their work that would coincide with the development of the first prototype engine.

But in order to have a bearings and seals program, there must be an engine design. The 5K team, until Jarlath McEntee was "borrowed for a few weeks" (it turned out to be a year) for the DOE heat pump project, had an abundance of design engineers, and they worked in three principal media, two of which were extremely high-tech and one of which was as old as Newton. At the team's disposal were a computer-aided design, or CAD, system and a very clever and patient computer draftsman, Todd Cale. This was one high-tech tool.

The other was the brainchild of David Berchowitz, a computer program called SAUCE, for "Stirling Analysis Utility Computer Emulation," a back-formed acronym in that David had originally called the program SOURCE, because it was going to be, in his estimation, the source of all Stirling-engine design for the rest of the century. After a few months of debugging, he scaled back his claims; instead of being the source of all good design, the program would be "like a little sauce, a little something extra that improves the final result."

SAUCE is an amazing tool, but, as David would be the first to admit, an imperfect one; rather than an invention, it was really a refinement and compilation of the work of many, including many at Sunpower or affiliated with it: David Gedeon's GLIMPS, or GLobally-IMPlicit Simulation program, Israel Urieli's early work at Witwatersrand, Costa Rallis's work there as well. (Costa J. Rallis, now professor emeritus at Wits, is a sort of godfather to Sunpower: David Berchowitz, Neill Lane, Nick van der Walt, and Izzy Urieli all studied with him there.)

SAUCE is incredibly useful for engine "optimization," or refinement of a prospective design, a tool that has almost the status of video game at Sunpower, in large part because it runs on the Macintosh computers so many engineers favor (for its user-friendliness), but also because it permits futzing around with engine design. SAUCE allows the designer to specify all of the parameters within which engine design occurs: temperature, pressure, diameter, mass and stroke of the piston and displacer, the characteristics of the heat exchangers and the regenerator; each SAUCE engine "run" generates a "loss inventory" that shows where the power goes: so much lost in friction, so much to hysteresis (or heat transfer to the wall of a cylinder), leakage through the seals. It calculates power output and efficiency.

The raw material for SAUCE comes in large part from the third source, the low-tech one: pages and pages of calculations done in heavy black binders assigned to each engineer and intended to provide a paper trail for every design. Supplementing the black binders are several texts, generally referred to by the colors of their covers: the "red books," six volumes containing tables and formulae for materials properties and entitled, appropriately, *The Metals Handbook,* and the "black book," which is full of actual test data for actual materials—strength and density and conductivity and stiffness and so on for all commercially made metals, plastics, resins, polyesters, and alloys.

Then there are the "orange binders," which are Sunpower-compiled primers on engine design consisting largely of memos from one Sunpower engineer to the rest announcing good ways of solving intractable problems; a half dozen copies of "David's Book," which is actually Berchowitz's dissertation on Stirling design from Wits; there are also the "notes," which are the Wits-trained engineers' actual class notes from school (these are kept, somewhat cryptically, in dozens of government surplus binders from the Securities and Exchange Commission, which makes Neill Lane's bookshelf look as though he were a tax lawyer—a very messy tax lawyer). Finally, there is *Roark,* as in *Roark's Formulas for Stress and Strain,* the bible of the stress-and-strain business, which provides a mechanical-engineering way of describing the Stirling-engine business.

Stress is the ratio of a force to an area in a given material; strain is the change in dimension of a material after it has been stressed. Stress is what you impose on a material when it is compressed or pulled or twisted; strain is what happens when a material is compressed, or pulled, or twisted. A piece of metal—a paper clip or a coat hanger—can be bent easily; the force one applies in bending a coat hanger is stress, so many pounds of pressure over the area, or diameter, of the wire in the hanger; gentle pressure can "flex" a coat hanger in such a way that it will spring back to its original shape; more forceful bending will change the hanger's shape irrevocably; the bending action is stress; the deformed coat hanger has been strained. *Roark* is full of "closed form" formulae covering thousands of configurations for calculating stress and strain, and it is thus indispensable to anyone designing a machine made out of metal that will be subject to force.

Roark is a fascinating book, full of little drawings of beams and plates and rods; the formulae that accompany the drawings are opaque to the layperson, but the pictures are not. A "case," or example, might be something like a beam supported at two ends: the formula shows how, given a particular material's characteristics, one may calculate the maximum force the beam will support before failing. The entire visible world is reducible to *Roark* formulae: why chairs have rungs, why wooden stepladders have metal rods underneath the steps, why some doors have two hinges and others three, why pianos have long, continuous piano hinges, and why elevator cables are woven out of many strands of wire, rather than being one solid piece. *Roark* offers a way of seeing everything in the context of stress and strain, although it is limited in certain circumstances peculiar to Stirling engines—Neill Lane says that *Roark* covers every case except those that you need in reality. For the world beyond *Roark,* Neill and Jarlath lobby hard for a "finite element analysis" computer program, although for nearly a year, it is not forthcoming. For the initial design, they make do without it.

So with a stack of books and a calculator, engine design begins; of course, every engineer at Sunpower also has the benefit of the experience of everyone else at Sunpower: no one at Sunpower is shy about offering suggestions to anyone about anything. But, all the same, the calculations must be done in order to get a rough sketch, a schematic, which must be done before SAUCE can be used. Only when a machine has been roughed out in this fashion can a part—the first of many, many parts—actually be designed, drawn, and fabricated.

The same process, in principle, is necessary to "design" anything— a doorknob, a dinner plate, or a drop cloth, on the one hand, and a space shuttle, a nuclear reactor, or a race car, on the other. Space shuttles and race cars are much more complex than drop cloths and doorknobs, but the activity of designing them is the same, involving Neill's "requirements, restraints, criteria."

Stirling engines are particularly devilish to design because they are dynamic machines that work at the limits of thermodynamics, areas where small changes in one place may have large effects in another. So Neill spent the first few weeks of design with a Macintosh computer running SAUCE set up next to a CAD machine running a drafting program called CADKey, a black Sunpower notebook on his lap, and, of course, his wrist in an itchy cast. He would make an educated guess at something—say, the cross-sectional area of one of the heat exchangers—and try it in SAUCE; then he would try something else.

Each iteration brought him closer to something he could actually envision in hardware, given the parameters he was working within. He knew the operating temperature, the maximum temperature for the cold heat exchanger (which was only "cold" in a relative sense; he was aiming for a cold side temperature of about 50 degrees Celsius, or 130 degrees Fahrenheit, not outlandish for desert conditions); he knew what the frequency of the engine should be (60 Hz), and he knew Power Out—five kilowatts, net, which meant he looked for something in the range of six to eight, in part because SAUCE tends, like David Berchowitz, to be optimistic and in part because the internal losses in the engine would quite likely be greater than predicted, at least until they were chased down in hardware.

When he had a rough idea of what he was in for, the team met with Robert Redlich, a soft-spoken and gentle electrical engineer who is perhaps Sunpower's most prized consultant; from Neill's rough impression of the engine, Robert began working on an alternator design: How many turns of copper wire would it need, and in how many coils? How many "windows," or current generating loops, would there be in the laminations? How many laminations would there be? What magnets of what strength would be necessary? Robert is an electrical engineer of the old school whose original area of expertise was in the arcana of antenna design and function. His knack for linear alternators is something that some engineers believe is God-given; whether or not this is so, linear-alternator design is at best problematic, in that the designer has none of the bags of tricks that conventional-alternator designers have, nor is there a century of experience and data to draw upon.

The early concerns with an alternator big enough to make six kilowatts or so of electricity have to do with heat and efficiency. Heat, always a by-product of making electricity, is an enemy of magnetization. Efficiency is an issue because the parts of the alternator—the magnet paddle and the inner and outer laminated cylinders of iron are hand-built structures, potted out of metal and glue and phenolic resin. As such, they will never be perfectly round, or perfectly concentric, and tiny variations in concentricity rob an alternator of power.

Size was an early issue as well: the larger the diameter of the alternator, the more power dense it would be, but there would be an upper limit to the diameter, for at least two reasons. The first had to do with physical size—the pressure vessel had some upper limit to size, as did the physical structures like the magnet paddle. The second limit had to do with the delicate meshing between the piston of the engine and the magnet paddle of the alternator.

In the 5K's immediate predecessor, the SHARP engine, the alternator and the piston were of almost identical diameters; while this made for an easy connection between the two—essentially, a ring of bolts—it limited flexibility when it came time for debugging and optimizing the engines. The SHARP engines were always underpowered, and Sunpower couldn't do the easiest power boost, which is to increase the diameter of the piston, because the size of the piston was strictly limited. So the engineers were left to try and squeeze a few watts here and a few watts there out of the many different internal losses.

On the other hand, an alternator with a diameter much larger than the engine's piston creates a physical problem of linking the two together. The alternator is essentially two thick, solid tubes with a narrow gap between them, in which the magnet paddle moves, sealed inside the pressure vessel. Connecting piston to magnet paddle was going to be a real trick, physically, without even worrying about power and efficiency. There didn't seem to be a way to hold the two parts together.

The problem, schematically, was like this. The magnet paddle was going to be about the size of a three-pound coffee can with its ends removed, although the walls were probably ten times as thick: a very sturdy coffee can. The piston was heading toward being about the size of the lid of a *one*-pound coffee can. The magnet paddle would weigh four or five times as much as the piston and would in essence be cantilevered out a long way, in terms of mass, from the piston. And the magnet paddle had to be perfectly concentric to the inner and outer lamination cylinders, so that it would pass through the narrow gap in the alternator without rubbing and not sag or angle downward. And yet it couldn't float in space: it had to rest on something, but it couldn't rest within the alternator, because it would rub and demagnetize the magnets.

What Neill hit upon looked plausible, at least on paper. Connect

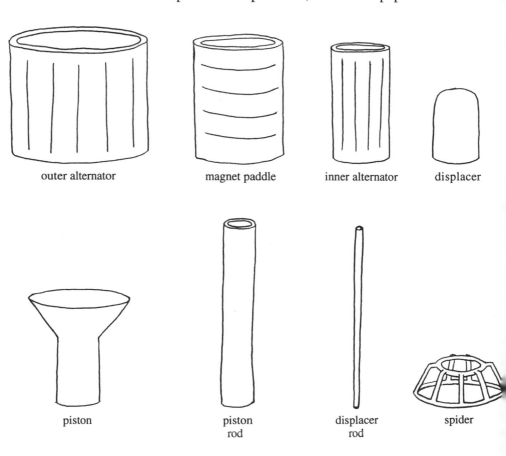

outer alternator   magnet paddle   inner alternator   displacer

piston   piston   displacer   spider
                          rod   rod

Drawing by Janet Cadmus

the piston to the magnet paddle with a pair of circular rings of metal of different diameters, the two rings connected to each other by metal legs. In engineering parlance, this is called a "spider" because it resembles a daddy longlegs looked at from above: small diameter on top, legs angling out to make a larger diameter ring at the tips of the spider's legs— a spider standing on a key ring. A bicycle wheel is spiderlike in this way, with many spokes for legs connecting the rim of the wheel to the hub. Through the center of the spider, Neill conceived of a long, hollow steel rod, small enough in diameter to pass through the hole in the center of the inner laminations, but large enough for the long, slender displacer rod to fit *inside* the piston rod, a rod within a rod within the alternator.

It looked something like a clever stacking up of household items. First comes the outer ring of the alternator about the size of a large bucket; then the three-pound coffee can–sized magnet paddle; then the inner ring of the alternator snugly inside, but hollow itself. Through this passes the piston rod, about the diameter of a cardboard poster mailing tube, and through this, the slender walking-stick-sized displacer rod. The piston face itself attaches to the mailing tube, and a metal spider connects the magnet paddle and the piston. The displacer rod runs through a hole in the piston face and connects at one end to the domelike displacer and, at the opposite end, to the gas spring puck.

But long before such parts could be fabricated, or even precisely drawn, they would have to be matched: the relationship between the engine and the alternator is crucial, because engine power is useless if it can't be gotten out to the real world via the alternator, and an alternator that isn't tuned to the engine will be at first inefficient and eventually nonfunctional: power created must go *somewhere*, and if it doesn't leave the engine/alternator gracefully through a pair of wires, it will stay inside in the form of heat, which will in a very short time cause whimpers or bangs. A whimper might be a hugely inefficient and hot alternator, wasting energy in the form of heat; a bang would be an engine that overpowers the alternator and runs away until it smashes itself into scrap.

The engine and alternator must be created in a perfect and multilayered balance. The engine makes the mechanical power; the alternator converts mechanical power into electrical power. A very efficient engine would turn a large percentage of the heat that powers it into mechanical power; a very efficient alternator would turn a large percentage of the mechanical power into electricity. So far, so good, but the two components must be matched in such a way that they are both at or near their peak efficiency together when they are at or near peak efficiency individually.

This is not automatic. Both engine efficiency and alternator efficiency increase with the length of the stroke of the engine's piston, which would tend to lead engine designers toward engines with large strokes. This would be a foolproof rule of thumb if a crucial component of the alternator—a ring of magnets—were not physically connected to the engine's piston, and thus responsible for some of the piston mass: the more mass on the piston, the more power required to move it. So in designing an engine, one wants as light a piston as possible; in designing an alternator, one wants as many magnets as possible, and more magnets means more mass. Hooking the two together means aiming for a combined efficiency that comes from taking advantages from the alternator and giving them to the piston, and vice versa.

It would be relatively simple to design an engine/alternator combination balanced in this fashion, if no other factors were involved. As one might suspect, however, more factors are involved, and changing any one has an effect—sometimes linear, sometimes exponential—on everything else.

A simple example might be something like this. In order to overcome the additional mass that the magnets necessary to the alternator's function add to the piston, one might increase the piston's diameter: a larger piston produces more power. But power—or, more particularly, the force exerted by the piston—increases by the square of the increase in diameter: a piston three times as large produces nine times the power, so relatively small increases in piston diameter makes for potentially large increases in power, and too much power can be just as bad, or worse, than too little.

Too much power can drive the piston with more force than the alternator can absorb, and the result then would be an engine that "runs away," the way that the pedals on a bicycle spin freely when the chain slips off and the rider keeps pedaling. On a bicycle, this may be disconcerting, even uncomfortable. In a Stirling engine, it is catastrophic, because the power has to go somewhere, even in an engine that is out of control. An engine that runs away will batter itself to death in seconds, as the piston and displacer collide, the rods bend or break, the magnet paddle deforms and bursts, the cylinders fracture. The vivid description of this around Sunpower is the image of tilting the pressure vessel and pouring out all the smashed-to-smithereens pieces of engine and alternator.

And this was but one of the fine points of the first axisymmetric design. In some cases, the team relied fairly closely on the designs and experiences of previous and current Sunpower projects: all Sunpower regenerators, for example, resemble one another closely, because their design has been refined through a dozen projects; the heater head on the 5K resembled the heater head on the SHARP engine, as well as the heater head on the DOE heat pump, because both represented the state of the art as far as Sunpower—or anyone else—could tell.

The bearings scheme resembled nothing that anyone had ever seen, and represented in its first incarnation all of the triumphs and failures of the design process. Neill Lane was quite proud of the initial plan for the bearings in the 5K because it provided a workable solution to a very difficult problem. As an ultimate solution, though, it was not the sort of thing an engineer would display at, say, a job interview: it was complicated, delicate, likely to be high in internal losses and short of life, and expensive to make. The first bearing scheme had "interim solution" written all over it, but it would certainly permit the design to progress through fabrication and testing and the first actual engine runs, during which time other solutions might become apparent.

The way the bearings were configured was this. The piston would be attached to a hard, hollow steel rod, about the diameter of a soda can, at one end, and at the other end would be the spider, attached to the magnet paddle. The rod would ride inside the hole in the inner laminations on a pair of hard steel sleeves called counterfaces, which would ride inside the actual bearing surfaces themselves; in the first configuration (for testing purposes only), the bearings would be machined out of bronze and coated with a Teflon-like material called Xylan. Xylan is a commercially available surface coating that is sprayed on like paint and that hardens into a hard but low-friction surface, like the inside of a skillet. The idea was that the hard steel counterfaces, ground to a slippery finish, would glide back and forth over the Xylan-painted surfaces with little friction and little wear. This whole scheme was duplicated inside the piston rod for the displacer rod: correspondingly smaller counterfaces and Xylaned bronze bearings.

Sunpower has used hundreds of gallons of Xylan over the years, for the same reasons that so many people use nonstick frying pans: it is in fact likely that Sunpower would no longer exist without Xylan, because the key to making machines that are efficient enough to compete in the real world is minimizing internal engine losses, of which friction is one and seal leakage is another. Metal rubbing on metal with no lubricant is disastrous in the way that frying an egg in a hot stainless steel frying pan without any butter is disastrous. And in addition to the bearings, the other crucial contact surfaces inside a Stirling are the seals around the piston and the displacer in their cylinders, because very small increases in these gaps have huge effects on engine performance. In the 5K, for example, the piston rides on carbon-fiber piston rings that fit tightly inside the cylinder; there is a gap, if one can call it that, of about two one-hundred-thousandths of an inch, between the rings and the cylinder. Yet through this gap escapes about 1,500 watts of power. If the gap were tighter, the loss of power might be less from leakage, but the loss to friction would increase, from about 500 watts to perhaps 1,000, offsetting any gain (these are good examples of internal losses, and how they are trade-offs, one loss for another). The tighter a seal, the higher the friction, and the greater the need for low-friction surfaces.

So Xylan coats the surfaces. Xylan is very useful from Sunpower's point of view because once it dries, it turns very hard, yet loses none of its low-friction properties. This means that a cylinder or a bearing surface coated with Xylan can be machined to a precise size and permit a very close fit. Xylan, which is a sort of cobalt blue in color, might well be Sunpower's motif inside its engines, the way that John Deere farm tractors are always green. There is lots of blue.

The final link in the bearings scheme for the 5K was a stroke of inspiration. A common bearing in the bearing industry is called a spherical ring bearing, a two-piece structure consisting of an outer ring and an inner ring, one inside the other, the inner ring with curved edges, so that a shaft passing through the inner ring can still spin freely even if not perfectly straight. They are as common as dirt in applications where shafts spin, and are inside all manner of mechanical equipment, from sewing machines to air compressors to drilling equipment to vacuum cleaners.

The principal feature that appealed to the 5K team was the ability of this type of bearing to accommodate an imperfectly straight shaft. For if the piston rod passed through two of these bearings (and, indeed, if there was another pair of bearings inside the piston rod for the displacer rod to ride in), the spherical bearings became, in 5K parlance, "self-aligning" bearings, permitting the piston rod, its counterfaces, and its bronze bearing surfaces to be in a straight line: the bearings have the ability to "absorb" small imperfections in straightness that are inherent in any long rod. The spherical bearings were first fitted into the inner alternator; then came the bronze bearing surfaces, then the hard steel counterfaces, and finally the piston rod itself. Inside the piston rod were a pair of spherical bearings, the displacer bearings, the displacer counterfaces, and the displacer rod. Normally, when such a large number of pieces are nested inside one another, their slight differences in size and trueness "stack up," or accumulate, as well.

For example, if each piece in the assembly had a tolerance, or acceptable deviation in size, of, say, a tenth of a millimeter (or four-thousandths of an inch, which is fairly large as machining skill and engine design go), and there were six pieces, the worst case might be if all the parts were a tenth too large, or six-tenths too large in toto. If the cumulative tolerance (for all six pieces) were supposed to be a tenth of a millimeter, then a tenth-of-a-millimeter deviation in each would be six times the acceptable tolerance. But the spherical bearings would compensate for at least some of this.

Clearances were so important (and need to be treated seriously in the design process) for two reasons. First, it is easy—too easy—to design a machine that "works" on paper but is impossible to fabricate—the type of material (aluminum, steel, stainless steel, cast iron), the size of the part (is it an inch in diameter, or six inches?), and the job it is to do (will it be hot? cold? load bearing?) all affect how closely a given part can fit in conjunction with another part. In a perfect world, all round parts are round and will stay round at any temperature or under any load: a piston 165 millimeters in diameter will be fabricated exactly 165 millimeters in diameter and will stay that way, no matter what.

In the real world, though, some tolerances are easy to hit, others are impossible. Aluminum is strong for its mass, but is prone to thermal expansion: it "grows" when heated. Stainless steel grows much less, but tends to deform under machining, so it can be difficult to make round; cast iron stays round, but is tough to machine to a precise tolerance. Further, the specification of tolerances is an art in itself. Imagine a piston that is supposed to fit inside a cylinder. Both will be made out of a material that can be machined to within, say, two ten-thousandths of an inch. If the piston is machined to four inches plus or minus two ten-thousandths and the cylinder is machined to plus or minus two ten-thousandths, there are six possible configurations that would fit the tolerance (cylinder plus two ten-thousandths, piston minus two-thousandths, cylinder minus two ten-thousandths, piston plus two ten-thousandths, and so on), but only *one* configuration in which the piston would fit inside the cylinder with two ten-thousandths clearance. A designer who doesn't keep all of this in mind ends up with parts that don't fit inside one another, or parts that fit together but then can't be separated, or parts that are too sloppy in fit to work.

A particularly clever—but permanent—trick with two pieces of the same material illustrates this. A rod and a ring of aluminum (a material prone to thermal expansion) machined in such a way as to have the rod slightly larger than the hole in the ring wouldn't seem to go together at room temperature. But put the rod in the freezer for a few hours, and heat the ring with a propane torch, and they will slide together easily; the rod will shrink and the ring will expand. When the parts reach ambient temperature, however, they will be inseparable: the rod will have expanded and the ring will have shrunk, and heating and cooling won't help separate them, because heating or cooling them will increase or decrease both, and aluminum conducts heat so well that it is impossible to heat one without heating the other.

And the clearances at Sunpower, while challenging, are not impossible to achieve, if the designers know what they're doing and are able to convey that knowledge to the machinist. But the clearances have to conform to the real world; they have to work both on the machine tools and in the machine, and the designer must never forget this.

Under the weight of time, money, and sponsor, it is tempting to forget this. The "appendix gap," for example (the tiny space between the displacer and its cylinder), is a Stirling engineer's temptation on the order of the apple to Eve: just a *tiny* decrease in the appendix gap can make an engine simulation on SAUCE look terrific: if Neill Lane changes the appendix gap in the simulation to zero, he gains nearly eight hundred watts of power, simply by closing a gap from *almost* nothing—one two-thousandth of a meter—to nothing. But "almost nothing" is possible, while no appendix gap at all is simply impossible. But it sure makes the numbers look good.

SAUCE, thus, can be either a designing tool or a video game. In the privacy of an engineer's cubicle, a bored designer can make an engine simulation of heroic qualities—copious amounts of power, undreamed-of efficiencies, loss free in every sense. Of course, the designer knows, deep inside, that such an engine can never exist in hardware, but it seems such a waste, say, to lose nearly a quarter of the heat transferred to the helium in the regenerator: while a nonengineer might marvel at the fact that the regenerator can capture more than three-quarters of the heat that passes through it, a video game–playing engineer isn't satisfied, and so increases the surface area of the regenerator in the SAUCE simulation and—presto!—gains a thousand watts of power. That such a regenerator cannot be fabricated is easy to forget.

When engineers play late-night video games on SAUCE, they are often learning something about the peculiarities of engine thermodynamics, and doing no harm; when overly optimistic projections creep into the real world, harm can be done, as when SAUCE simulations are used as a basis for selling Sunpower's expertise and capabilities. Someone who has not spent several hundred hours working with SAUCE is not likely to understand the esoterica; what he or she sees instead are the numbers like "power to load," which is what Sunpower sells, and "efficiency," which is how well Sunpower can design an engine to deliver power to load. If non-SAUCEans don't read the fine print, they will be disappointed at least, outraged (or litigious) at most.

But when the 5K team actually begins to fabricate parts, to make the pieces of metal small and large that go into what would be the real engine, it is not always easy to tell whether someone is "futzing"—improving something that is good enough—or fixing something that won't otherwise work. It is enough to make the engineers on the team tetchy themselves—with each other, with the sponsor, and, as the parts come out of the shop and begin to be assembled, with the engine itself.

The transition occurs when the group can no longer work only in the abstract—in the drafting room or with the SAUCE simulation, or in hours and hours of consultations and conversations and meetings. The balance shifts into the tangible world when there are enough inanimate engine parts to begin putting together, and, as Neill says late one afternoon, "the *real* problems start."

Chapter 8

Shade

We told NASA [on the Space Stirling program] that if it didn't work, no problem: we'd just send Shade up there on the space shuttle, and he'd make it work.

—*Neill Lane, on the skills of Test Engineer David Shade*

HE IS A HIGH SCHOOL GRADUATE from New Jersey who learned to be a machinist from his father; at age eight, Dave Shade could already run a milling machine. His loves are loud music, fine tolerances, junk food, clever ideas, bad jokes, hot coffee, conspiracy theories. On Halloween, he dresses up as one of the Blues Brothers and holds, on a twenty-foot string, a jack-o'-lantern-printed orange plastic garbage bag filled with helium. He is building a magnificent house in the country, a house he designed himself, a house built to the specifications of the California earthquake building code. It will be a solid house, solid like Dave Shade.

He is strong, which goes without saying, because all machinists are strong, but gracefully so; his hands are sturdy like a bricklayer's, but his fingers are deft, like a cellist's. He can tighten a dozen bolts within a few ounces of torque, one from another, all by feel. In a machine shop with some very good machinists, like Jerry Royse and Ray Klinebriel, Dave Shade has no peer. He says with disdain (it's so obvious) that if you can draw a part, you can make the part.

Such certainty is exploited regularly by engineers like Neill Lane, who when he must design a part that looks, on paper, impossible to make—a large diameter with a tiny tolerance, a long rod with a gentle taper, an aluminum cylinder so thin that it changes shape when carried from a warm room into a cool one—Neill will say, mournfully, to Shade that Jerry has looked over the drawing and said that such a part simply can't be done. "Lemme see the drawing," says Dave Shade.

"Hell, I can do that." And he then does so. Dave Shade loves to make things work. He *makes* things work. At strategy and task assignment meetings (which he never attends—"Why should I sit around in there and listen to people *talk?*"), everyone wants Dave Shade on his team, because there is no one remotely like Dave Shade at Sunpower, or perhaps anywhere else in the world. He is resolutely and completely unflusterable. "Shade," Neill Lane will say as he watches Shade do something like delicately trim a ten-thousandth of an inch from the diameter of a piston, "is simply incredible."

Even to an unpracticed eye, the confidence that Dave Shade exudes in manner, style, circumstance, and skill is apparent. See him climb a ladder up, up, up, to the high ceiling of the machine shop, safety glasses propped on his forehead, a heavy black box of a circuit tester hanging around his neck from a strap, the wire leads of the tester in each hand. At the top, he pulls open a junction box and casually puts the tester leads across a 240-volt, 60-ampere fuse: see, in other words, Dave Shade fifteen feet off the ground on a metal extension ladder, poking around in a pre–World War II rat's nest of wires that contain the current and voltage of an electric chair. He loops one arm around a piece of conduit, clips one of the leads to the junction box, and raises one knee to prop up the circuit tester so that he can read it. He then executes a particularly tricky maneuver: he flicks an inch-long ash from his cigarette. "You got current," he calls down to Jerry Royse, whose milling machine has inexplicably stopped, necessitating a climb to the ceiling of the machine shop. Shade reaches into the junction box with his sausage-like fingers and wiggles the fuse, causing the engineers gathered around the base of the ladder to close their eyes and electrical inspectors across the Midwest to wince. "Don't electrocute yourself until you finish my part," calls Neill Lane. Shade smirks, closes the box, and double steps down the ladder. "That's nothing," he says. "You should have seen this shop I worked at in New Jersey; there wasn't a fuse in the place, just all these bare wires." To Dave Shade, life is simply a continuum of dangerous conditions, poor equipment, recalcitrant machine tools, impossible deadlines, wrongheaded designs, ill-executed plans, and those-who-forget-the-past-are-condemned-to-repeat-it decisions, all in need of attention from someone who *really* knows what he's doing. Anyone who meets Dave Shade knows that he knows what he's doing.

That Shade should have ended up at Sunpower seems inevitable (it is virtually impossible to imagine the place without him, given the high percentage of dangerous conditions, poor equipment, wrongheaded designs, etc., etc.), but in fact it was simply accidental—small companies in Ohio don't generally recruit machinists in the New Jersey shipyards. Rather, Shade and his wife were just passing through Athens one summer on their way home to New Jersey, and they thought Athens was a quaint little town, the sort of place they would like to live. So they up and moved, arriving the same day that Ohio University's seventeen thousand students returned from summer vacation. "Uh-oh," Dave said to himself. "What have we done?" What they had done was move to a town with one large industry—a university—that needed few machinists, even twenty-four-year-old machinists with sixteen years of experience. Someone suggested he try the banjo factory, which turned out to be as unlikely a place for employment as it sounds: "In this town, anyone who's run a drill press calls himself a machinist," the banjo factory said, but they did send him up the street to Sunpower, where he has been since January of 1982. It was a remarkable stroke of good fortune, because Stirling engines—mysterious, balky, un-derfunded, quixotically conceived—need Dave Shade.

There was a time not long ago in America when machine tools—lathes and mills and saws and presses and grinders used to turn, cut, saw, drill, and surface the metal components that are used to make more machines—were fundamental to the national economy. It was with machine tools that virtually everything that makes something else was made: the auto industry, for example, is a large collection of men and women operating machines that make parts, which other men and women (and, more and more often, machines) assemble.

Washing machines, aircraft engines, bicycle frames, hot-dog rotis-series, lingerie, comic books, and nuclear weapons all have their origins at some point in a pair of machine tools—the milling machine and the lathe. A milling machine cuts ("mills") holes and shapes with a rotating bit much like a drill bit; a lathe uses a tiny metal blade to cut and shape a piece of rapidly rotating metal into a cylinder, a tube, a rod. The advantages of a machine tool over, say, an electric drill and a saw are many, but two are the most important. First, machine tools are made to work with large and heavy materials—Sunpower's two largest lathes can be used to machine a steel disk as large in diameter as an automobile tire, and three or four times as thick, a piece of steel that might weigh hundreds of pounds. Second, machine tools are necessary for precision, which, in addition to his exceptional ability to visualize a complicated machining process, is Dave Shade's stock-in-trade.

Stirling-engine designs such as one finds at Sunpower are exqui-sitely precise, designs created just at the cusp of what is feasible to do.

Sometimes this feasibility lies at the upper limit of what it is possible to do given a particular material and a particular machine tool and a particular machinist, the difference between doable and reproducible. A doable design is one that someone can execute; a reproducible design is one that *anyone* with widely available tools and readily available skills can execute. The difference is huge, because while there are many machinists, there is only one Dave Shade.

So a set of standards and practices have evolved, immensely complex standards, governing specifications (the shape of a part), tolerances (how close to the specification must the part be), dimensions (the measurements of a part), materials (tolerances, for example, are different for different materials, depending on their composition and inherent qualities, often referred to as "machinability"), indication (the act of aligning material in a machine tool), order of processes (referred to as the "set up" of a job and a tool), and on and on and on. Machinists' manuals are written like the Ten Commandments, full of "shalls" and "shall nots" that all good machinists know; the best ones know them at an instinctual level, having as they do the ability to envision the finished part within a chunk of featureless metal, like a sculptor. Neill Lane, when he is watching Shade make something, often quotes the "how-to" of sculpting an elephant: "You just cut away everything that doesn't look like an elephant." This is what machinists do. Some have learned in trade and technical schools, but most have in fact learned on the job, starting with sweeping up metal chips and ending up running their own machines.

The aspiration for a machinist is to make a part that resembles exactly the drawing by which it was specified. Bad machinists are the ones who make all manner of modifications and "improvements" to jobs; machinists and machine shops are known by their reputation for how close they come to the ideal replication of a drawing. In fact, there is a curious "ethic" in machining about which there is always some discussion, and by which one can get to know a machinist's personality. Since tolerances are typically given as a range above and below the "exact" measurement, some machinists work in a manner such that their setup and operation is simply to produce parts that fall somewhere within the tolerance, while others aim for the exact dimension, using the tolerance as insurance against miscalculation. A third strategy is to go beyond machining to actually envisioning how a part is to be used, and determine a tolerancing scheme based on that: is one tolerance crucial, while another is clearly less significant?

So, three different machinists making the same part, say, an aluminum cylinder with an inside dimension of two inches, "plus or minus five-thousandths," and an outside dimension of two and a half inches, plus or minus fifty-thousandths, might employ three entirely different gestalts. One machinist aims for *exactly* two and two and half inches; another for two inches plus two and a half or three- or four-thousandths (less material to cut away) and two and a half inches plus forty- or forty-five-thousandths (ditto); a third machinist might aim for an exact dimension on the inside, because the tolerance suggests a close-fitting part, but a sloppy tolerance on the outside, because the tolerance suggests that it is not particularly important.

But to call Dave Shade a "machinist" is something like saying a sculptor makes marble chips: he may be a very good machinist, but, by definition, being a machinist has an upper limit to mastery. What Shade also has, as no one at Sunpower except perhaps William and David Berchowitz has to the same degree, is an intuitive sense for what makes Stirling engines go, an ineluctable entwinement not with the Stirling cycle, or thermodynamics, but with the hardware itself. It is Shade who, more than anyone else on the 5K, makes the machine come alive.

It is not exactly true, however, that a million-dollar project is balanced on the shoulders of a high school educated machinist from New Jersey; it just seems that way, some days. Shade does nothing unique, unless having such a wide range of skills and experience is unique; there is nothing he does that *no one* else can do, but he can do more things more cleverly and skillfully than most, and is able to do so with a minimum of direction. There are technicians (and Sunpower has on occasion hired many of them) who would spend a day fruitlessly trying to fit tab A into slot B because they were told to do so, and they have no faculties for judging why it might not fit. Shade knows not only how but why; he is confident enough in his skills to dispute, to suggest, to modify, so that something works, though without violating the spirit of the original intentions.

This experience and judgment have been hard won by Shade, in more than a decade of working on recalcitrant machines at Sunpower. His memory of what has worked, what hasn't, what has been tried, and what the result was is encyclopedic; many a time, he begins a suggestion with a phrase like "Here's what we tried on SHARP, only it didn't work because. . . ." And then he proposes the correction. When the SPIKE engine was sent to its sponsors, Kawasaki Heavy Industries, in Japan, Shade was sent along with it; his one phrase in Japanese, he always insists, is "Where is the vodka?" Of all the technological wonders of the Orient, he professes to have been most impressed by vending machines that dispensed whiskey—"right there on the street—how conveeeenient!"

This is another side of Shade, hard-living and hard to know when to take seriously. He is blunt-spoken to what some might consider a fault. "Well, that's a bullshit idea," he said once in a design review; another time, he pointed at a place on a design drawing and said, "That's so *stupid*." He is perfectly capable, with a straight face and a look of shock, of telling Neill or Abhijit that some crucial part has just been destroyed, and he can stick with such a line far beyond what might normally be considered a practical joke; after three months on the 5K, Beej's absolutely reflexive response to anything that Shade says becomes, "No, c'mon, really? Be serious." When Shade was in Japan, and a technician inadvertently left a helium charge line on SPIKE in the Kawasaki test cell and the pressure vessel failed, Shade's telex back to Sunpower read, "Engine blew up. Send another," a telex the sponsors quickly amended to "Pressure vessel clamp failed; send another." Only Shade would think of joking via telex.

Shade's sense of humor—of delight—is a constant source of entertainment, all the more pleasing because he has the ability to joke, to be funny, *and* to work. Sometimes he seems simply to be amusing himself; he drives a Renault Reliance, but calls it an Appliance; he writes out "distructions" for the machinists; when strong backs gather to move a heavy piece of metal, he suggests filling it with helium. His favorite tool, he maintains, is duct tape; on his toolbox is a novelty button that reads, "I ♥ Anesthesia."

Some of his best lines are all the better for being completely private: he is pleased to amuse himself. One summer, William's daughter, Faith, undertook to arrange a nice historical display in the old glass cases in the tiny lobby of Sunpower, gathering the first free-piston Stirling engine, the first free-piston water pump, one of the earliest linear-alternator machines, and so on. She laid them all out on the table in the conference room, William wrote notes describing what each bit of hardware was, and Faith set out to make neat and impressive labels for everything.

Shade happened by while she was doing this, and he went immediately to the morgue to root around. What he found was a cardboard mock-up of an idea William had had once for a resonant pump, a mock-up held together with Shade's favorite tool—duct tape—at the edges of its oddly pyramidal shape. It looked exactly like what it was, a half hour or so's worth of doodling with cardboard and tape and hose clamps that was then abandoned, but never thrown out. (Sunpower saves everything.) Shade contemplated this bit of detritus, and then, when Faith was out for lunch, he put it on the conference table next to the more substantial bits of Sunpower hardware and scrawled a note to place under it. Six years later, in the big oaken display cases in Sunpower's lobby, near the first FPSE, still sits this little cardboard pyramid, neatly labeled from the information Shade thoughtfully supplied on his note, the " 'Cheops'-type Resonant Pump." Apparently, no one has noticed it, or thought much about it, except for Dave Shade and perhaps Larry Nelson, who once in a blue moon carefully, reverently dusts the objects in the case, including the "Cheops"-type resonant pump. It is one of the funniest things Dave Shade feels he has ever done, although—or perhaps because—no one else is in on the joke.

It is the mark of an active mind, a mind not often in repose. When he is working—machining, fabricating, assembling—he can keep up a conversation that suggests catholic tastes in reading: a UN report on the status of women in the Third World; military technology; philosophy; motorcycles; constitutional law; pharmaceuticals ("Did you know that thousands of anesthesiologists are addicted to inhalation anesthetics?"); the specifications of aircraft. And, of course, Stirling engines. Shade collects stories of Stirling failures the way others collect moments of success, because of a kinship with William that Shade seldom acknowledges, he too is rooted in hardware, in making things work. A comprehensive awareness of which things *don't* work, then, is a tool as important to him as any other. Stirling engines are strange and unusual, and it interests Shade to work on something unusual; he wouldn't be happy making washing machines. A foreign visitor to Sunpower once (and this may have been a problem in translation), managed to ask Dave Shade where one put the "fuel" into a Stirling engine, and for weeks Shade considered mocking up a gas cap of some sort onto the pressure vessel, a coup almost as resonant as the Cheops pump.

A prototypical Shade design is an interesting vision to contemplate as the 5K begins to become an engine. Shade's prejudices are for tried-and-true solutions—not necessarily elegant, but solid. He dislikes O-rings, for example, the slender rubber gaskets that the world associates with the space shuttle *Challenger,* for *Challenger*-like reasons: for O-rings to work, the grooves in which they rest need to be exquisitely machined; O-rings that rest in grooves on outside diameters—like a rubber band around a coffee can—can be tricky when it comes to assembly. To watch him watch Abhijit make an O-ring of a nonstandard size can be hilarious; chin resting on hand, he has a look that says, "No *way*." (Abhijit learned years ago to ignore the looks he gets when he does anything involving tools or manual dexterity.)

Shade is an advocate of the centering scheme called "centerports" as a method for causing the piston to run in its most advantageous place in its cylinder: *all* Stirlings eventually need centerports, he will point out, so why not put them in in the first place? He despises "mated" parts, as when a long rod is made up of two or more sections and then threaded together; they are never true. One of the earliest components of the 5K design was a "kill valve," intended to provide a foolproof way of stopping the engine if trouble arose: the plan was an electrically controlled solenoid valve that would "deflate" the displacer gas spring and stop the engine. From the moment Shade first saw the kill valve, it became the object of his derision. Unnecessarily complicated, tricky, undersized, it was not a Shadian solution. A *real* kill valve, Shade would say, would contain plastic explosive. It is hard to know when to take him seriously.

As the subassemblies of the 5K begin to come together, the skill of Dave Shade becomes not only apparent but necessary. A subassembly is the sum of its constituent parts and, in theory, *simply* the sum of its constituent parts. In practice, conglomerates of parts take on shapes other than those their designer intended, and must be "tweaked" to within spec. The outer laminations of the alternator need, one afternoon, to be tweaked. The outer laminations of the alternator weigh about a hundred pounds and consist of thousands of thin sheets of iron glued into a concentric cylinder, just larger than the inner laminations plus the thickness of the magnet paddle. The inner surfaces of the laminations themselves are bare iron; between the lamination faces are hundreds of windings of copper wire, coated with a special nonelectrically conductive epoxy. The whole assembly, then, is a big, heavy cylinder of iron and copper wire and glue and flexible rings of phenolic resin, and it was supposed to be, in the language of a machinist, "dead nuts" on in concentricity: a perfect cylinder made from imperfect materials. A lack of concentricity would make for a place where the magnet paddle would rub; a rub on the magnet paddle would, at best, ruin it; at worst, it would ruin *everything,* by damaging the alternator in such a way that it couldn't absorb the power produced by the engine, and the whole thing might well blow up. Tilt the pressure vessel and pour the parts out.

So with a hand from Ray Klinebriel, and with Neill smoking cigarettes and hovering, Dave Shade hoists the outer lamination assembly into a lathe; for a moment or two, several months and many thousands of dollars of work are hanging by a steel cable, swinging gently at a forty-five-degree angle. "*There*'s the problem," says Shade, and Neill asks, anxiously, "What?" Something *else?* he wonders.

"The whole thing is made on a forty-five-degree angle," says Shade, deadpan.

"Dave . . . ," says Neill, shaking his head; there are times when he thinks Shade will drive him round the bend. Shade just grins and guides the laminations into the "chuck" of the lathe on which he works most frequently—"Shade's lathe," it is called, a 1960s-vintage Colchester fabricated largely by hand and with exquisite care in England. "The World Turns on Colchester Lathes," reads the nameplate.

The chuck jaws on a lathe resemble, on a much larger scale, the chuck that holds the bit in an electric drill; for lathes, there are "inside" jaws and "outside" jaws, depending on where the material is to be clamped; further, there are "three-jaw" chucks and "four-jaw" chucks, suitable for different purposes—three-jaw chucks have three clamplike jaws that operate in concert with each other and that can be tightened and loosened simultaneously: with certain materials and in certain situations, they are "self-indicating"; all three jaws move together and thus clamp a piece of metal in their center, something like the way a camera tripod levels a camera. Four-jaw chucks are trickier, in that each jaw is independent of the others; a four-jaw chuck is used when the material to be lathed, or "turned," is irregular and when both the inside and the outside of a cylinder, for example, need to be cut in order to make the cylinder concentric. For even the simplest task on a lathe, Neill Lane's elephant-sculpturing analogy is applicable: a machinist needs to "see" the perfectly concentric cylinder that lies within a nonconcentric part, and cut from inside and outside accordingly.

To cut the laminations, though, adds several complicating steps. First, there is no real "true" surface on a potted-up-glue-and-iron-and-copper-and-phenolic pseudo-cylinder: what does it mean to say "straight" or "concentric" in relation to an object where no two surfaces are straight or concentric in relation to each other? (Imagine a deck of playing cards glued together with varying amounts of glue between each card, and with the outside edges of the cards not quite lined up: the edges and top and bottom are all "straight" in and of themselves, but not *square* in relation to one another.) Further, as everyone knows (but as Neill continually reminds Shade), the lamination faces need to be cut, but the covering over the copper windings cannot be; it simply cannot: bare copper wire would mean a "dead short" in the alternator, a place for electricity to arc across the air gap and cause a series of electrical events, all catastrophic and all uncontrollable. Electricity following its proper and intended path is safe and docile, the way water inside properly connected pipes is; but a "leak" of electricity causes much more trouble than a leak of water, if only because water can ruin your wallpaper or your floor, but electricity can kill you.

A "leak" of electricity between the windings in the alternator can also kill, or at least grievously injure, the entire 5K project; the disastrous months and money spent on the Box alternator hang over every move that anyone makes, including the moves that Dave Shade will make on the outer laminations.

Once the lamination assembly is grasped firmly in the jaws of the lathe chuck, Shade begins to "indicate" the part in the lathe. Indication is a process, not a gesture, a series of steps to adjust the part so that its physical location is perfectly perpendicular to the lathe's cutting tool. If it isn't perpendicular, the surface will end up tapered. If one envisions coring an apple, it is easy to see that if the knife is held at an angle, the core will be cut out at an angle as well; it could also be off center. Finding the center of a cylinder is not trivial though: as with, say, a doughnut, the center is in empty space.

So lathes are designed themselves to be stout and solid and straight, and the machinist adjusts the part by manipulating the jaws of the chuck and taking measurements with a device called (appropriately) an "indicator," moving the part, not the tool. (Lathes are very heavy and precisely sited and leveled in a machine shop. To "move the lathe" is a nonsense task assigned to apprentice machinists who don't know that lathes are anchored with long, thick bolts directly into concrete.)

An indicator is a spring-loaded rod about the size and shape of a ballpoint pen attached to an indicating dial. Rotating the part, by hand, in the lathe with the indicator point resting against it, will move the needle on the dial and show the machinist whether the part is "running out," "running in," or, what he wants, which is dead nuts. High-quality hard steel in a three-jaw chuck is easy to get centered exactly in the lathe. The outer laminations will be—Shade and Neill know this before they start—literally impossible to center, because there is no true surface on the assembly. Since, however, the steel rings at either end of the assembly were true when they themselves were fabricated, one ring is clamped in the chuck, and Shade indicates off the other. The first time he rotates the laminations against the point of the indicator, the dial runs out crazily and then back in, twice as crazily: each revolution of the needle is a thousandth of an inch, which to a precision machinist is a huge number; a ten-thousandth of an inch is still big: a good machinist with good material in a good lathe can cut one *ten*-thousandth of an inch away. Shade reaches for the chuck key, with which one adjusts the chuck jaws, and also for a large mallet, the machinist's favorite tool. When Shade taps the laminations with the mallet, Neill winces as though he had been rapped on the thumb. "Dave . . . ," he says, but Shade is unflappable. "Don't *worry*," he says. "What's the worst that can happen?" The question makes Neill wince again.

Shade taps, and then spins, taps and then spins; when he has the indicator needle spinning within five-thousandths, he moves the indicator point to the front, or face, of the laminations, and spins the part again; again, the dial rotates crazily. Having gotten the part near center on the lathe, he now must adjust it so that it is perpendicular as well. Tap, spin, tap, spin. He moves the indicator again, and the part runs out almost as much as when he started. "It ain't round," Shade says conversationally.

"I *know* it isn't round," Neill starts to protest, before he realizes that Shade is kidding him. Tap, spin, tap, spin.

Each iteration brings the part closer to where it must be, and Shade stops rapping with the mallet and instead uses delicate force on the chuck jaws themselves, always tightening with the chuck key ever so gently; it doesn't take much force to move something a thousandth of an inch. Press, spin, press, spin; move the indicator, press, spin, press, spin, press, spin. The outer laminations have not been moved enough to see with the naked eye, but the whole part has been coaxed by Dave Shade closer and closer to center and perpendicular. This is not a rare process—it is what machinists do—but it is a delicate one, and unless someone has been trained and given lots of practice, it can be an opaque process as well. Neill and Abhijit both know what Shade is doing, but as they watch him, they both also know that he is doing something that they cannot.

A subtle and mostly unspoken tension exists between those trained academically—an engineer—and those trained technically, or practically—a machinist, a technician—in every place where inventions are conceived and designs executed. Some engineers are frankly insensitive, considering technicians to be a sort of living tool, a wrench with legs. These are the engineers who curtly assign tasks without any indication as to why something should be done and who, in doing so, make for themselves the troubles they squarely blame on technicians.

Likewise, some technicians are reflexively disdainful of book learning and theory when it is not accompanied by any sort of *practical* talent or skill. There are technicians at Sunpower who revel in knowing that they can do things that the engineers would have no hope of being able to accomplish, particularly when it comes to the use of tools, especially hand tools, since many college-educated engineers have no feel whatsoever for hand tools. A fair number of the engineers at Sunpower are clumsy to the point of ridiculousness, especially those who are not mechanical engineers and thus haven't had even the rudimentary training with tools that accompanies an engineering education. Once, Eric Bakeman (history major turned computer scientist) asked Hans Zwahlen for a "star-shaped" screwdriver, which even for an engineer is extreme; Hans, like the other technicians, has watched engineers use micrometers for glue clamps and torque wrenches as hammers, but this request was frankly stupefying. In perfect confirmation of Hans's disdain, it turned out that Eric wanted a Phillips screwdriver to poke a hole in something. Whenever Abhijit whipped out his Swiss army knife (the ultimate engineer's tool, pretty and clever and elegantly designed, but not a *real* tool), Hans and Shade would rib him mercilessly; when faced with a huge bolt to be loosened, they would call for the "right tool for the job"—Beej's knife.

Serving as bridges between theory and practice are men like Dave Shade, with immense practical knowledge, an excellent understanding of theory, and wide experience, but no engineering degree, or Robi Unger, titled as an engineer by Sunpower, but not formally trained as one, or Al Schubert, working toward an electrical-engineering degree at Ohio University, but working as a test engineer at Sunpower. They have had a long practical apprenticeship that is in other countries a crucial part of engineering education but that in this country is seldom considered as "important" as a foundation in mathematics and theory. That Shade and Robi and Al all understand the levels of complexity involved in designing a Stirling machine helps them and the company in hundreds of ways, of course. That they are not, strictly speaking, engineers occurs to most people, if it occurs to anyone at all, when they do something that engineers can't, rather than when they do something that engineers can.

Most engineering students spend a month or a semester in a machine shop "learning" about as much about machining as premedical students "learn" about surgery when they watch a gallbladder being removed. Joerg Seume, an engineer on the 5K team, is an exception; educated in Germany, he spent nineteen months working in the machine shops of Daimler-Benz, apprenticed to a master machinist, an apprenticeship that begins with a file and a block of metal that the students, over weeks, turn into a variety of shapes. It breeds respect for materials and tools, and also for machinists.

All of the Sunpower engineers know the theory of machining; some of them have done it. If Neill or Abhijit were alone in the building and had to machine a part to save their lives, they could do it, but in practice Neill doesn't even pretend: what would take Shade five minutes might take Neill an hour, and they both know it. (When Abhijit makes the slightest move to do anything that looks like work in the machine shop, all the machinists stand protectively in front of their favorite machines, which leaves Abhijit with the worst lathe in the place. "C'mon," Shade said to Beej once, blocking the path to his beloved Colchester, "this is a nice lathe.") Some technicians are sensitive to anything they perceive as a slight or as arrogance, becoming themselves reverse snobs about skills—one technician used to shoo engineers away from certain tasks with a sense of glee, as though they were children "playing" at a *real* job, that of hardware and tools. If the vast majority of the engineers are guilty of anything, though, it is of bending over backwards to avoid any hint of condescension to the technicians and machinists, an attitude that has a collateral and happy result: this stance makes for particularly thorough and complete instructions, which makes for better execution and fewer mistakes. Giving such instructions is never easy—a half dozen times a week, a machinist will troop through the offices looking for a particular engineer, drawing, calipers, and part in hand, to say, "Are you sure this is what you want?" In these situations, engineers—or most of them—almost reflexively apologize for being confusing, reminded as they are that a part isn't done until it exists in hardware. Once, Shade came into the drafting room looking for Janet Halbirt, one of the draftspersons. "Is Janet here?" he asked, holding up a drawing. She was not. "Well," said Shade on his way out, "I'll just have to *guess.*"

Complicating a process already fraught with possibilities for error, for miscommunication, is that, astonishingly, and in the face of the most rudimentary and obvious common sense, the engineers at Sunpower use one system of measurements and the machinists use another, entirely different system. When a missed decimal point or the transposition of a pair of numbers can render hours of work useless, it seems more than passingly odd that Sunpower engineers work exclusively in the metric system and that Sunpower machinists fifty feet away work exclusively in the English system, the system of inches. The one exception is the company's newest milling machine, a Taiwanese knockoff of the famed Bridgeport mill it stands next to; when the Taiwanese machine was manufactured, one of its adjusting devices was mistakenly calibrated in millimeters, rather than in inches like the rest of the ma-

chine and, indeed, the rest of the machines in the shop. It is an error
that symbolizes the dichotomy between meters and feet that exists
at Sunpower.

The computer drafting machines on which the design drawings
are made place neat metric dimensions on the drawings; the machines
in the shop that are used to execute the designs are scaled and calibrated
in the English system. So the first thing a machinist does with a drawing
is sit down with a calculator and laboriously divide every dimension on
the drawing by 25.4, in order to convert it to inches, and then write the
dimension on the drawing above the metric dimension. Engineers
think in tenths and hundredths of a millimeter, and in microns; machinists in thousandths of an inch. A tenth of a millimeter is four-thousandths of an inch; a ten-thousandth of an inch is 250 microns.
Machinists say, easily, "a half a thousandth" and "forty-thousandths,"
meaning inches; engineers say, easily, "a tenth" or "a half a tenth,"
meaning millimeters. When engineers and machinists talk to each
other, they spend a lot of time saying ". . . *inches,*" and hearing *"millimeters"* in response, and vice versa; they also spend a lot of time doing
quick conversions in their heads. But all of the 5K parts are conceived
and designed in millimeters and then fabricated in inches. Oddest of all
is that most machines—drafting and machining—have digital convert-
ers that permit one to move from one measuring system to the other at
the press of a button, but engineers cannot "see" thousandths of an
inch, and machinists can't "see" 250 microns, and so they are always
dividing or multiplying by 25.4, a number that is the lingua franca of
the dimensional world. It can seem extraordinarily confusing, especially
when Sunpower ships a part to an outside shop for some process, such
as the grinding of hardened steel, that its own machine shop cannot
accommodate. Because Sunpower's drawings are done only with metric
dimensions, the parts end up depending on some stranger's dexterity,
not only with a machine tool, but with a calculator as well.

When Shade has the part indicated on the lathe—not dead nuts
but as close as he can be—he and Neill engage in a spirited colloquy,
joined by Abhijit, as to how best to proceed. The idea is to run the
cutting tool inside the laminations and first trim off any high spots that
are egregiously high—"bumps"—and then use the cutting tool to trim
off a fine layer of the laminations themselves. Abhijit wants the minimal
cuts taken at the slowest possible speed; he is worried about exposing
the copper; Neill is willing to go a little further; he is worried about
leaving something to rub on the magnet paddle. Shade wants to try and
take as clean a cut as he can, in order to make the inside of the cylinder as

concentric as possible. When he turns on the lathe, and the laminations begin rotating, Beej immediately says, "Whoa, whoa, whoa," and Shade stops the lathe, puts his hands on his hips, and says, "I didn't *do* anything yet."

"That's awful fast, isn't it?" Beej says, referring to the rotational speed of the lathe. This is relative; faster speeds, depending on the material, are generally better, because they permit a smoother, more controllable cut, and the tool will be less likely to hang or catch and jerk the part out of the lathe. On the other hand, going slow permits the machinist to watch very carefully and take very shallow cuts, seeing the point at which he ceases to cut away high spots and starts cutting on a continuous surface. Beej is worried about cutting into the copper; Shade is worried about jerking the part out of the lathe and onto the floor; Neill is worried about both, one more than the other, evidenced by his Solomonic decision: "As slow as you *can*, Dave, but as fast as you need to." Shade first switches on the lathe at a snail-like speed—a revolution or so every fifteen seconds. "How's that?" Then he lights a cigarette, adjusts the lathe speed up to where he wants it (and where he had it in the first place), checks his tool, clears the digital readout, leans forward, and guides the tool into the maw of the laminations. Neill and Abhijit both turn away, so as not to seem to hover, but strain to watch out of the corners of their eyes. Shade ignores them; he was running a machine tool when Beej and Neill were playing in the family garden. The tool touches and makes a *screech* sound, followed by silence, then a screech, silence, in a regular rhythm; Shade is getting the high spots on the first row of laminations, and the tool whines when it is cutting, but is, of course, silent, when it is not.

There is a certain naturalness that some people have when handling instruments with which they have an affinity; it might seem highfalutin to compare Dave Shade to a concert violinist (his musical tastes run to extraordinarily loud rock and roll), but it is hard to watch him concentrate on a task without thinking of the word "virtuoso." He stands at the lathe, one hand on the tool control, leaning over and peering into the spinning laminations assembly, and every fifth or so revolution his eyes glance up at the digital readout and then back down; it is the stance of someone who knows exactly what he is doing at a level of mastery measured in microns—or ten-thousandths of inches. He stops the lathe, and Neill and Abhijit both lean over to look inside the laminations as Shade assembles a "bore gauge" of the proper size to measure the inside diameter of the assembly. A bore gauge resembles an adjustable-length metal rod that can be locked, with a twist, into any length; Shade touches it across the inside of the laminations and then measures the length of the rod with a large set of micrometers; he counts, looks up at the ceiling (converting from inches to millimeters), and then looks at the drawing. He puts the bore gauge back inside and measures again, in a different place. Neill can't stand it. "Well?"

"It's damn close," says Shade. "It was, what? about eight out of round on the faces before, right?" And he waits while Neill does *his* conversion, this time eight-thousandths of an inch into millimeters. "It was that much out of round, yes?" says Neill.

"Right," says Shade. "Okay, now it's about two, but the whole thing is . . . four bigger."

What Shade has done is impressive, even for Shade; whereas the insides of the laminations were, prior to tweaking, anywhere from zero to eight-thousandths of an inch thicker, they are now not only four-thousandths of an inch thinner (effectively making the hole in the doughnut that much larger), the concentricity varies by only 25 percent as much—a double win. And he did it without hitting the copper. Beej runs his fingers gingerly inside the laminations, feeling the surface. "So, let's get it out of there, ay?" And to his misery—Shade can drive someone *crazy*—Shade loops one of his huge forearms underneath the laminations, reaches up with his other hand to loosen the chuck, and smartly hoists the assembly out of the lathe. He stands there holding it, a grin of amusement on his face that neither Neill nor Abhijit can see, because they both have their eyes tightly closed. He sets the laminations on a work table and says, "You can look now. *Geez.*"

"Dave . . . ," says Neill, shaking his head for the umpteenth time this day.

"Lighten up," says Shade, grinning and lighting another Kool. "What's the worst that could happen?"

Beej shepherds the laminations back to the assembly cell in the lab, and they break for lunch—Beej to eat impressively healthy-looking sandwiches ("Another bean sandwich today, Beej?" asks Shade), Neill to uptown Athens, Dave Shade to consume an impressively unhealthy-looking pair of sausage hoagies at his workbench, reading, this day, the UN *Report on the Status of Women in the Third World*. When they reconvene later in the conference room, Shade simply asks, "What next?"

There is the milling to be done on the transition piece; ("Doc is working on that on the afternoon shift," says Shade. "It's dumb for me to maybe mess up his work"); there is the displacer rod to be assembled and trued, the displacer dome to be assembled and trued, the bearings debugged and recoated with Teflon-like Xylan.

Then there is the heater head, the displacer gas spring, the cooler shell to be finished, the cooler itself leak tested. Also the piston and displacer rings cut, the backup regenerator finished, the alternator assembled. Also the feed throughs for control and monitoring need to be parsed out, tested, and debugged. ("No one has *touched* the electronic stuff yet," Shade points out.) They need to figure out how to assemble the various subassemblies, and how to assemble the subassemblies into an engine.

By the time they are through, three-quarters of an hour later, Abhijit has a neat—and very long—list in his notebook of things to do. They prioritize, and assign tasks. Watching Abhijit neatly date the page in his notebook, Neill realizes that they are doing this on the very day when, according to what can most easily be referred to as "the original revision of the original revised schedule," they would be collecting a payment from Cummins as a result of having run the engine for four hours. Just to torture himself a little bit more, Neill then points out that according to the *original* original schedule, they would be fast approaching the day they would deliver to Cummins a five-kilowatt solar Stirling engine. That is, the *second* one. "Well," says Shade, "at least the pressure's off for *that*."

In a sense, he is right. But this only means that the pressure has for all practical purposes doubled for the one engine they do have, or hope to have. Shade heads to the machine shop, Abhijit to the telephone to call John Bean at Cummins, and Neill Lane to the coffee pot, after they have agreed to reconvene when they have all worked through their lists. "See you guys in about a month, then," says Shade, "and we can make *another* list."

Chapter 9

More Heat, More Pressure

There are literally dozens of things we can do with the bonus money.

—*William Beale, following the contract signing with Cummins Power Generation, Inc.*

THE RELATIONSHIP BETWEEN Sunpower and its sponsors is something like a relationship between two people. The vocabulary is similar, as are some of the dynamics. Sunpower courts sponsors all the time, for example: gets dressed up, puts on its best clothes (all the engineers except William own at least one necktie), plans the equivalent of a first date. In addition, every few weeks or so, John Crawford, the senior engineering manager, ends up on what might be thought of as a blind date— someone from some corporation who has heard about Sunpower, perhaps from a friend of a friend, who calls up and wants to talk about the possibilities. In this fashion, John Crawford (and whatever colleagues at Sunpower are appropriate) may meet with manufacturers of industrial pumps, middlemen in the Third World water pump business, makers of lawn mowers, vaccine coolers, truck engines, vacuum cleaners. Once or twice a year, Sunpower is asked out by a crackpot interested in perpetual motion.

While all this is going on, Sunpower is not sitting at home waiting for the phone to ring: if there is a potential sponsor in the world that has not received at least one phone call or letter from Sunpower, it is an oversight plain and simple. John Crawford is fond of mailing lists, sending out batches of a hundred or so technical papers, letters, or fliers about Sunpower to, say, companies that might have some use for the small engine; Lyn Bowman, Sunpower's electrical engineer and electronics maven, is of the disposition to simply call some large corporation and ask to speak to someone there who has an interest in certain technologies and has money to spend. (While this may seem unusual in the extreme, it is in fact exactly how Sunpower landed one of its largest sponsors for the small Stirling cooler they call the Cucumber.)

Of all these first dates, though, only a few result in second or third meetings; of those, fewer still result in any sort of relationship at all. Sometimes this is because Sunpower is picky; most of the time, it is because the prospective sponsor is: Sunpower appeals only to a certain sort of suitor, one willing and able to pay development costs, which start at six figures and go up quickly to seven.

Once upon a time in the industrial world, paying for research and development was considered part of doing business—every large corporation had its own R & D division, and many corporations had several. Such internal R & D programs were encouraged to stretch limits, to expand the envelope.

But what was once very common is now very rare. The reasons for this are complex, but the net result is that most large corporations don't like to spend money on something for which neither success nor customer demand is assured. As has been pointed out to Sunpower time and time again, if Sunpower were simply *selling* something like the small engine in quantities of, say, ten thousand units, lots of people would buy them, and many companies would be delighted to play middleman in the process. But R & D always means that the first prototype costs ten or fifty times what the ten thousandth item off the assembly line will cost—a small engine that might retail for a thousand dollars must begin life as a $10,000 prototype, which in itself might cost $50,000 or $100,000 to develop. Sunpower is not a cheap date.

At the beginning of the 5K program, Cummins Power Generation, Inc., seemed to be the ideal partner for a long and pleasant relationship with Sunpower. For one thing, CPG was itself something of an anomaly, a division of a large corporation that was formed specifically for R & D. The Cummins Engine Company saw in the early 1980s the potential for alternative-energy research—the company's main product line, diesel engines, is mature to the point of being gray haired, and Cummins foresaw a day when there might be a call for alternatives to nineteenth-century technologies, with their thirst for fossil fuels and their environmentally unfriendly exhausts. One division of CPG is investigating "clean" combustion engine technologies; another is pursuing diesel power plants for the utility industry. And one CPG division is pursuing solar thermal power generation with the Stirling engine.

CPG entered into its relationship with Sunpower with its eyes at least half open; CPG's president, Jerome Davis, knew he was taking something of a gamble but, at the same time, thought that solar thermal was a good match for Cummins's expertise in several visionary ways. Cummins was a good manufacturer, for one thing—the diesel engine business is so cutthroat that only good manufacturers can stay in business. Cummins was the largest independent maker of diesel engines in a field where it is not enough to sell, say, truck engines to truck companies: it is a convention of the trucking industry that individual customers—indeed, individual owner-operators—select which engine they prefer, a frightening state of affairs for any company that doesn't know exactly what it is doing. It is as though instead of buying, say, a Honda automobile or a Ford, one could pick the Honda's transmission, a Chrysler engine, a Toyota air conditioner, and a Ford suspension. Truckers and trucking companies demand this every day, so a company like Cummins cannot simply sell thousands of engines. It must sell each engine, one by one. This makes for a rather aggressive approach to manufacturing and marketing.

And what Cummins saw—perhaps the first time that anyone in the private sector other than Sunpower saw this—was the inherent potential of solar thermal power conversion as an opportunity to sell thousands, even hundreds of thousands, of something that a company could make. Such opportunities are rarer and rarer in the contemporary world, the chance to make something out of big pieces of metal and sell it in very large numbers. The personal computer industry is more typical of present-day manufacturing and sales opportunities, in that computers get smaller and smaller and cheaper and cheaper with every new generation. To be in on what might be the beginning of an industry that might result in selling many megawatts' worth of Stirling engines manufactured by Cummins was an opportunity that CPG was willing to take a chance on, even to the tune of an initial investment of a million dollars. Cummins, like many large American corporations, also knew how to play certain tunes with government money—funding from the feds for research and development that required a private-sector partner.

Indeed, it was Cummins's assessment of the market potential that caused a lot of the shifting and jockeying at the beginning of the 5K program—an eight-kilowatt engine? a five-kilowatt engine? a pair of four-kilowatt engines?—because Cummins wanted to cover all the bases. In the major leagues of electricity generation, five kilowatts is very small—almost nothing: a small conventional power plant produces ten megawatts, or ten million watts, and a large conventional power plant might produce a gigawatt—a billion watts. Because of the infrastructure required to build a power plant, economies of scale dictate that bigger is better; as with Stirling engines, the first megawatt is the most expensive, while the tenth megawatt of capacity is relatively cheap. The five-kilowatt size was not exactly arbitrary—it matched handsomely with Cummins's market research for a much smaller market, that of remote electrification and water pumping—but it was also never assumed at the beginning that the 5K would be the ultimate machine. The experience of designing it and making it work would help dictate the size and configuration of the ultimate machine, which was thought to be somewhere around twenty-five kilowatts. Forty 25Ks would make a megawatt, which is about the smallest number that power conversion experts can envision working with, and forty of anything sounds easy to envision. That no one had ever produced a single watt (until the Sunpower/Cummins SHARP proof-of-concept program in 1990) of electricity by means of this technology was a stumbling block, of course.

But solar thermal energy conversion has never been a field for either the timid, the poor, or the easily frustrated. The first hurdle for solar energy has long been the automatic association by the uninformed between the phrase "solar power" and tiny photovoltaic cells such as one finds on pocket calculators. Nonengineers hear "solar power" and think of a technology that has nothing to do with solar thermal conversion, but which does fit the popular conception, and they can be forgiven for not being able to imagine how such a wimpy little plate of glass could provide enough electricity to run a city the size of Los Angeles. The open mind necessary to understand the fundamentals of solar thermal conversion is fairly rare: the nonengineer hears "solar" and thinks "sunlight"; the engineer hears "solar" and thinks "heat," heat being the currency of engineering.

When prospective sponsors visit Sunpower, they expect to see some device outside, running on the heat of the sun. The company, after all, is *Sun*power. In the time it takes to explain the second law of thermodynamics to a nonengineer (a rule of thumb at Sunpower is that the more money an individual controls, the less likely that individual is to be an engineer), valuable attention-span time is lost; visitors are puzzled and often too polite to say so. This is compounded by the use of electric or natural gas–fired heaters to test and debug Sunpower engines; these "heat pipes" are large, expensive, and often contrary pieces of equipment that make Stirling engines look more complicated and toggled up than they really are, even at the prototype stage.

Until fairly recently, when Sunpower adopted the practice of using photographs of the SHARP prototype on the solar concentrator in Texas as an introductory visual aid, lots of visitors left Athens frankly suspicious: they saw no sun power. (Sunpower, for all its years of work on solar thermal, has only recently had a solar dish in Athens at all, a clever modification of a TV satellite dish up on the roof that William installed for future testing of the small engine. Although William finds it baffling, the simple existence of this dish is comforting to those unconversant with the technology; it is invariably one of the first things that visitors see, and comment upon, when they pull through the gate on Byard Street. There's a *solar dish,* they say. We must be in the right place.)

CPG was aware of this from the beginning, and as big corporations tend to be, it is lavish with visual aids: photographs of the solar concentrator, schematic drawings of the concentrator and engine, and so on. Its initial brochure on the 5K project was notable for consisting mostly of photographs.

The other obstacles to solar thermal have long been more real than apparent, though. They are primarily two: the mirrors, which focus the sun's rays; and the conversion system, which turns heat into electricity. The former have traditionally been extremely expensive and delicate, and the latter is no different from—and thus no cheaper than—that used in conventionally fueled power plants. All things being equal, solar thermal was always likely to be expensive when it came to capital expenditures.

The dozen or so solar thermal conversion power plants around the world are testimony to this. The most common configuration is a tower in the middle of dozens of concentric circles of fragile and exquisitely ground glass mirrors, all precisely placed to focus the sun's rays on a vessel filled with some absorptive fluid, like water. The mirrors are so numerous because the sun, of course, moves across the sky, and a mirror that is perfectly focused at nine in the morning is out of focus by nine-forty-five. The mirrors themselves must be ground in the same way that expensive telescope mirrors are ground—often by the hands of skilled craftsmen. A nick, a bubble, a rough spot, and the mirror's efficiency drops precipitously. A sandstorm can damage mirrors; a hailstorm can destroy several million dollars worth of them (and months of work) in a few minutes.

And all of this is finally a way of boiling water, which is used to run steam turbines that make electricity. While such installations are environmentally sound, they are difficult to think of as economically sound—the sale of electricity is a business of tenths and hundredths of pennies (it is not some accountant's perversity that calculates electricity rates out to three and four decimal places: it is part of the business), and as long as coal and oil and natural gas and nuclear power are available and acceptable to a society, conventional solar thermal will always have a hard time competing: the "fuel" may be free, but the means of procuring it is not.

The 5K program goes to the heart of these two difficulties, representing philosophical as well as technical differences. The mirrors used for the solar concentrator dish on the 5K program are nothing short of astonishing when compared with the ground-glass mirrors they replace; the "mirrors" are in fact not mirrors at all, but two sheets of shiny plastic film supported on an iron ring like a Hula-Hoop. The film looks exactly like that used to make shiny party balloons, although it is slightly shinier, slightly thicker, and only slightly more expensive. And rather than one large mirror, the 5K concentrator consists of twenty-four mirror facets, each about five feet in diameter. The results, when compared with that of conventional glass mirrors, are profound. The plastic film is light, shockingly cheap, and durable; glass mirrors are heavy, expensive, and fragile. The plastic film facets are focused by simply vacuuming the air out from between the two sheets; a small vacuum pump draws the film into the concave shape necessary to focus the sun's rays, and an electronic control system makes it possible to defocus the mirrors in a few seconds—important in cases of trouble. Because the mirror facets themselves are so light, the whole concentrator system is light, and easily moved by an electric motor to track the sun, so placement of the solar concentrator can be accomplished by a technician with a longitude and latitude chart and a compass—by contrast, the permanent glass mirror installations require precise placement of each of hundreds of mirrors. What is lost in reflectivity when compared with glass mirrors (about 80 percent versus 93 percent) is made up fiftyfold in cost savings, site preparation, and maintenance. It is hard to overstate how fundamental a difference this is: at "SolarOne," Cummins's solar dish test facility in Warner Springs, California, the company has fabricated and installed six hundred solar concentrators of the type and size necessary to run 5K engines: if it had six hundred engines, it would have a three-megawatt solar installation, and would have it for pennies on the dollar compared with the cost of glass mirror solar power.

The other profound change from conventional solar thermal to Stirling solar is how the electricity is actually produced. All electricity, no matter the source of the energy—coal, nuclear fission, natural gas—is produced, finally, by a generator made of magnets and copper wire and steel laminations. The Rancho Seco nuclear power plant, in California, which produced a gigawatt of electricity, and the little bicycle lights that work by means of a knob rubbing against the tire of a bicycle both work in the same way. Solar thermal plants that use the sun's heat to make steam are conceptually and practically like Rancho Seco: big and expensive installations. Stirling solar thermal is more like the bicycle light: relatively cheap and relatively small, because instead of one huge generator, there are many, many smaller ones, and one can buy and add capacity in very small increments. A company, a municipality, a cooperative, could buy electricity in small and inexpensive increments, rather than all at once.

The development of a cheap and flexible solar concentrator and discrete, small generation systems changes the face of solar thermal profoundly; in conjunction with a Stirling engine, the whole strategy of solar thermal changes. Whereas before, any solar thermal installation was sized in the manner that conventional generating plants were sized—on the order of at least a megawatt, or one million watts—and priced accordingly, the 5K program presented the possibility of many small generating plants, of generating plants that could be built incrementally to meet demand, of small installations for places where it would be extraordinarily expensive to provide electricity in any other way—remote sites, for example. And rather than selling a multimillion- or multibillion-dollar power plant every five or ten or fifteen years (which is not good for cash flow), Cummins foresaw selling thousands of units every year, the way it sells diesel engines.

Indeed, the first half dozen customers for the 5K prototypes speak to this potential. Two of them are what R & D companies call "soft targets"—big and well-funded operations that are not real consumers, but instead are technology incubators: the South Coast Air Quality Management District, or SCAQMD, in southern California, and the Electric Power Research Institute, in Palo Alto. Both are buying 5K systems not to make and sell electricity but to test the potential for the technology.

But the other customers are downright scary in their real

worldness: Georgia Power, Pacific Gas & Electric, the Pennsylvania Energy Office, and AT&T (which has research facilities in remote and sunny places). Granted, they are buying units for testing only—all are careful to stress that they make no commitments beyond setting up a dish and running it; and granted, the customers within the customers are R & D units themselves. And granted, all these "customers" are only doing the field testing of prototypes—no one is selling kilowatts of Stirling solar power yet. But no one at Sunpower can help thinking ahead a little bit, dreaming a little bit, when the news comes that a real electric company will be making real electricity from a Sunpower engine. The first customer is the hardest. To read the letters of commitment from companies like Pacific Gas & Electric (the quintessential big power company, supplying as it does the most energy-hungry region in the United States—northern California), though, it is easy to dream. They have committed themselves to installing and operating a Solar 5K system—PG&E, in fact, is going to get *four* 5K systems. All Neill Lane and the 5K team need to do is design and fabricate the first one, and make it work. Since everything must flow from this event, making the engine and making it work takes on rather indomitable proportions around Sunpower.

Some days it seems inimitable as well, a terrifying thought to Neill Lane; as the design moves from calculation and simulation into the actual fabrication of hardware, the scope and scale of the project become clearer and clearer to him, and the result is an inevitable realization that the success—or failure—of the 5K program will ultimately depend on many people, a fair number of whom he can see and talk to, but also a fair number of whom he cannot. (Some parts must be shipped out to specialized machine shops for certain procedures, for example, and three important subsystems—the solar concentrator, the cooling system, and the electronic controls—are contracted out to companies other than Sunpower.) The project and the multitudinous skills required to bring it off are simply too immense, the steps involved too many, for one person to pretend to have a grasp—let alone *control*—of the whole project. The team has always relied to a certain degree on teamwork; at some point, it becomes a group effort, people working on discrete tasks at different levels of intensity, but not exactly working together. Rather than something like a boat in which everyone rows in unison, the 5K begins to resemble an ocean liner, where there are many people doing things that the captain can't even see, let alone do himself. Neill can't do everything—he never thought he could—while at the same time he knows that every task is a point where the whole project can collapse. Outwardly, he remains philosophical; he keeps his temper; he tells Abhijit again and again when problems crop up that there are going to be problems; he tries to be methodical and reasonable. But it gets harder and harder to do as they invest more and more in the project.

An analogy that comes to Neill one evening is not one that he needs. In ostensible off-duty relaxation, he turns on the local PBS television channel and watches a documentary that follows the progress of the design and building of a large Manhattan office tower. Viewers who were not at that moment in charge of large, complex projects fraught with the potential for disaster might have been amused at the difficulties, some minor, some major, that would be expected in such a big job. One type of material is ordered, but another type is delivered. A last-minute change is instituted only partly, because the instructions weren't disseminated to everyone who needed to know. A major decision is made without the consultation of the person who knew the most about the ramifications of the decision.

For Neill, watching the show was a harsh form of torture. "The first thing I see is this interview with the guy who's supposed to supply the bricks for the building," Neill recalls, "and he's saying in this know-it-all tone that the architect doesn't want the bricks he ordered, but really needs this other kind of brick, and I *immediately* think of William, who's *always* making some *huge* change and saying that it's what the sponsor needs; this guy is standing in some brickyard hundreds of miles from the building—from a building he's never *seen*—and saying, 'Oh, I know what they need better than they do.' And then back at the site, someone gets the bright idea to change the place where the elevator machinery will go, and they build *half the building* before someone thinks to do the calculations, and it turns out that they can't put the machinery in the new place after all—the building won't support the weight. So they have a half-finished office tower with no place for the elevators. By now, sweat is *pouring* from my brow—this is supposed to be my *relaxation* and it's giving me an ulcer; I feel like I'm going to be sent to engineering hell for this disaster, and all I'm doing is watching a television show. I'm half watching and half running through every weak spot in the 5K program in my mind when they show this big group of workers who are supposed to begin some part of the project—*hundreds of workers*—and they cut to the steel yard that was supposed to supply the steel for these guys to assemble, and it's all piled up there half a continent away because it was never shipped. I thought I was going to pass out before I could turn the set off, and then of course I didn't sleep a wink for about three days. And when I *did* sleep, of course I'm dreaming about the engine; I can't *wait* to get to work in the morning so that I can *do* something, instead of just brooding about it."

As the actual engine parts begin to take shape, Neill and Abhijit fight such demons daily, if not hourly. The original 5K solar II design—the axisymmetric design—was a compilation of sorts of the different ways that Sunpower had conceived of solving various problems. Part of this had to do with the brutal time constraints caused by the Box alternator effort; part had to do with the consequent financial constraints caused likewise. The rest came about, paradoxically almost, because of Neill's determination to make the 5K project qualitatively different from all other Sunpower projects, in that it would proceed in a methodical fashion no matter the pressures of time and money, because Neill believed that this was the *only* chance the 5K had to get out the door; conventional wisdom at Sunpower had long held that the trick to R & D was to get something that worked. The 5K team believed that the trick was getting something that someone else could *use*. It is an important distinction. Take, for example, the bicycle spokes.

To understand the bicycle spokes—there are fifty-six of them in the 5K—one has to understand why Neill Lane loves them, and why William Beale hates them, and how and why they—the bicycle spokes—work. It is also a way of understanding Sunpower, and how it often works.

What Neill Lane loves about the bicycle spokes is simple: they perform a crucial function simply and easily. That function is to provide a particularly dicey form of strength and support to the magnet paddle, in that they act as a sort of backbone, connecting two thin metal rings at either end of the paddle. The spoke and ring assembly looks something like a cylindrical birdcage open at either end; the magnets are glued in between the spokes in rows of three and then wrapped with Kevlar thread (the sort of thread used in bulletproof vests). Most of what holds the magnets in place is the glue; what gives the magnet paddle strength and, more important, concentricity are the bicycle spokes. The thread holds the magnets in place in one plane, keeping the paddle from collapsing at the glue joints. The spokes hold the magnet paddle's shape in the other plane, giving the magnet paddle far more rigidity than one would get from the magnets and glue alone. The spokes solve the problem that thwarted the attempts at making the flat magnet paddle, with the added advantage of supporting the goal of making the paddle round and concentric. The magnet paddle was assembled and glued, over two days and with much fussing, around a solid steel "mandrel," or form; when the glue was dry, the mandrel was pushed out one end, and the spokes and Kevlar and glue held it together. Voilà. There was a nerve-racking half an hour or so at the hydraulic press when the mandrel wouldn't come out (it turned out they were pushing at the wrong end, against a shoulder on the mandrel that was larger than the magnet paddle itself, an event that resulted in what may be a new indoor record in creative cursing), but when the paddle finally came off, it was exceptionally round, fairly concentric, and, most important, quite solid. The spokes worked.

What William Beale disliked about the bicycle spokes was not exactly clear in the beginning. On the one hand, they seemed to fit well with his philosophy of trying to use easily obtained parts and materials, real-world stuff, to make Stirling engines. On the other hand, he complained several times about their very existence as well as their use—once, he stood in the drafting room and bent one of the spokes that were lying around back and forth, absently, while speaking with Neill about something else, as though he were tearing up a paper clip like a desk jockey. Even after the magnet paddle was assembled, he asked one more time, "So you're using those bicycle spokes?" and then he laughed his dismissive laugh, as though it were one of the more harebrained schemes that anyone had come up with.

A clue to this distaste came a little later in the fabrication process in a slightly different context. Neill and Abhijit and Dave Shade and Jerry Royse and one of the machinists, Doc Licht, were deep in consultation around one of the milling machines, upon which rested the single largest and heaviest part of the 5K: the "transition piece," an awkwardly named and oddly shaped slab of steel about the size of an automobile wheel. The transition piece served several functions, all of which contributed to its shape and machining. One was to serve as a cap for the open end of the pressure vessel; another was to provide the surface to which the heater head was attached (in this sense, it is the "transition" between the pressure vessel and the heater head and heat pipe, hence its name). The transition piece had holes through it for the electric and electronic feed throughs—where the power would come out—and a large hole in the center in which the piston cylinder sat. William came by, hands in his pockets, and listened to the discussion for a little while and then wandered away. The transition piece seemed to be coming in for a lot of unnecessary attention, he observed. "They're spending too much time making the part *look* nice," he said; "it's a damn prototype, and they're fooling around with a lot of extra machining. A prototype should just be big pieces of metal—who cares what it looks like—and they're doing a lot of work that they don't need to be doing now." He laughed his little laugh then, too.

To William, "prototype" has a particular and precise meaning: a prototype is the first working version of an engine, made for the sole purpose of validating the principles of its design. It is not pretty; it is not necessarily durable or long-lived; it is not final, but by definition ephemeral. Any effort spent in making a prototype anything more than this was effort taken away from other tasks: learn what you can from the first quick-and-dirty engine, and then make all your refinements.

So, the bicycle spokes: the bicycle spokes didn't solve an immediate problem—getting enough parts close enough to what they ultimately might be so that an engine will run—but instead solved a "luxurious" problem, that of making a durable and long-lived magnet paddle, one that would last not for five or ten hours but for five hundred or a thousand. A day spent machining the transition piece that would make it useful for a hundred-hour run was time taken from the first run; the several days spent figuring out how to make a magnet paddle with bicycle spokes was time spent fussing, rather than machine making. The 5K team was awfully fussy; "no one expects that any of the parts in the first engine will be the same as in the second or third engine," William said, "so why spend any more time than is absolutely necessary to getting the thing running?"

But Neill wanted (if not expected) parts in the first engine to be as close to what would be in an ultimate machine as was possible, because he was not depending on providence or fortune for a second or third attempt: too many prototypes were in the Sunpower morgue, or in storage rooms in sponsors' warehouses, machines that worked, but weren't worked *through*. "Figuring it out in the second engine" was something the 5K team members were damned they would do unless it was something entirely unforeseen. Any problem they encountered they would solve, or try to solve, as it came up. But this absorbed time, and money, and energy, and patience. And it made them look fussy.

To some degree, they were being fussy, no doubt about it; there was some good-natured kidding, and some not so good-natured grousing, on the part of other Sunpower engineers over the 5K team's use of drafting and machine shop time when other teams needed these resources, and some evident amusement at Neill and Abhijit's insistence on certain protocols for assembly and subassembly procedures. In many circumstances, the 5K team would produce not only the drawings necessary for the fabrication of a part but also drawings for the fabrication of whatever assembly jigs or tools the part might require. When the jig was somewhat complicated, this seemed perfectly appropriate; when the "jig" was simply, say, a simple cylinder of aluminum used to guide one part into another, it seemed less so—Sunpower had long relied on institutional memory and ingenuity alone for such tools, Dave Shade and Al Schubert in particular being legendary for looking at a pair of parts and either remembering what had been used to put them together or figuring out something that would work. To make computer drafting drawings and assign part numbers to dozens of jigs and mandrels and rigs and then to carefully label (often in Abhijit's neat block printing) each of these pieces of metal seemed, if not fussy, at least ostentatious. One afternoon, Jarlath McEntee spotted a piece of metal in the 5K assembly cell that looked about perfect for something he needed on the heat pump project. Outwardly, it looked pretty much like many hundreds of odds and ends of steel lying around Sunpower, a round, solid chunk of carbon steel about the size of an institutional can of soup. To order one from a supplier would take three or four days; here was one that could be had immediately, so he asked Neill if he could "borrow" it, and Neill said, as politely as he could, no.

"What do you need it for?" Jarlath asked, such borrowings being absolutely commonplace around Sunpower.

"It's a mandrel for the magnet paddle," said Neill.

"So when are you going to make another magnet paddle?" asked Jarlath.

"Next year, but that's not the point."

"Well, what's the bloody point, then?"

"Jarlath, I'm sorry, but we can't." This was the sort of exchange that made the 5K team seem so fussy.

Abhijit's rationale, supported by Neill and, for that matter (at least in theory), by everyone, was clear-cut: they were not simply to deliver an engine to Cummins; they were to deliver, first, a reproducible engine and, second, everything that would be necessary to reproduce it—a sort of engine "kit." Institutional memory was fine when it worked, but the downside—institutional Alzheimer's—was needlessly risky. There were enough cold-sweat things to fear as it was without losing some part or process between the cracks and then having, when the pressure was really on, to improvise or prevaricate. Be straightforward, but be assertive: 5K stuff was 5K stuff, and don't put us on the spot by asking for it.

As assembly progressed, though, there was a subsidiary reason for this, one that for any given chunk of metal wasn't necessarily crucial but that could add up. Abhijit not only managed a lot of the details for the project; he also "paid" the bills, by billing to the 5K project and, hence, to Cummins every hour of time, every bolt, every chunk of metal, that they used. Consumables like machine tool bits and shop rags and epoxy glue and cleaning fluid were built into every project's overhead, but materials for actual parts were billed through to the sponsor—in an accounting sense, the sponsors "owned" everything they had been billed for whether it was used or not. Sunpower had learned this the very hard way on some government contracts when the auditors had shown up and asked to see such and such a piece of metal. (To this day, the U.S. Army owns a goodly heap of metal stock that is locked in a cage in the Sunpower machine shop; most of it is of little use or interest, but occasionally Sunpower engineers look wistfully at some perfect piece of material inside the cage. But they can't have it; they're not even supposed to "borrow" it. If and when the U.S. government goes belly up and sells off its assets, some liquidator can come to Sunpower and auction off the junk in the Army cage. Until then, it rusts.)

But Abhijit wasn't worried about the government; he was worried about the budget for the project. If one ignores the Box alternator, they were famously ahead of schedule and under budget; with the money and time spent on the Box, though, they were behind on both, and Cummins was a bit grouchy. Sunpower wasn't grouchy so much as anxious about the specter of its running out of money. And while of course any given chunk of metal wouldn't make a bit of difference in the budget for a million-dollar project (steel, or at least most garden-variety kinds of steel, is astonishingly cheap—pennies per pound; shipping often costs more than the material itself), Beej was girding himself for demonstrating how careful and efficient Sunpower had been with Cummins's money, if it came to that sort of showdown. Such showdowns had happened before in various forms and guises at Sunpower. With government contracts, there was not so much a showdown as simply a grinding to a halt. The government would quit paying the bills, and that was that. With much more latitude in contract negotiation involving private sponsors, it was not unheard of for a sponsor to end up in William's office bartering, cajoling, or demanding, with William bartering, cajoling, and demanding right back. It is one of the perquisites that sponsors get which Sunpower begrudges more than anything, this right to lord it over Sunpower simply because the sponsor is paying the bills. And money at Sunpower was getting tight, in general, and on the 5K program, in particular.

Money at Sunpower is a curious thing by real-world standards. It is obtained in chunks that sound large, but never turn out to be so. Some of this is due merely to accounting sleights of hand, because everyone with business sense knows that you hold on to money for as long as you can; a standard contract with Sunpower might involve the issuing of a once-monthly purchase order for engineering and design work valued at so many thousands of dollars; Sunpower then sends an invoice for that many thousands of dollars; the company then pays the bill by sending a check. This keeps most of the money in the sponsor's account for a long time, and very little money in Sunpower's account at any given time, because Sunpower, in essence, puts the money up in front by paying its employees, its own bills, and so on, and then getting reimbursed when the purchase order comes. (Sunpower is always trying to cajole sponsors into putting the purchase order in the overnight mail; sponsors generally send it via parcel post.)

Another reason that big money turns small at Sunpower is the sheer overhead of running the place. In addition to the fairly typical overhead of salaries, insurance, rent, and the like, there is overhead peculiar to R & D: machine tool parts and maintenance, a staggering electric bill that William is always trying to whittle down by turning off lights, esoteric and expensive testing and monitoring equipment, and the expense of trying to figure out what is going on in the thermodynamic universe, to evaluate it and monitor it and understand it. At its most flush, Sunpower employed, for example, a mathematician and two computer experts who spent most of their time modifying and updating the computer simulation and data acquisition programs that permit Sunpower to function. While at any given time their working hours are billed to individual programs, not all of their time can be; they are part of the overhead. So, too, are things like computer software and hardware, which pose a typical R & D conundrum: buy them outright and absorb the cost in overhead, or bill them to a sponsor, so that the sponsor pays, but is in an accounting sense the owner? When Jarlath and Gong Chen were asked by the Department of Energy to do a particular computer analysis for the DOE heat pump, the necessary software was procured through DOE, to save Sunpower the expense. The paperwork took three months; the analysis took only two. On the other hand, when Jarlath and Neill agitated for a finite-element-analysis program (which they insisted no engineering company could possibly do without), Sunpower bought it outright, and it was delivered the next day—but at a cost of several thousand dollars, which Sunpower had to ante up out of overhead.

These are not unique costs of doing business, of course—every business, from a bakery to a bomb factory, deals with them all the time. Sunpower is somewhat handicapped in that it doesn't have anything to sell—to generate cash flow—and is further handicapped by working in a field notorious for being expensive (this could be said about R & D as well as about Stirling engines). But in the matter of the 5K, Sunpower was handicapped by something that has handicapped it many times over the years: optimism.

When CPG and Sunpower set out to negotiate a contract for the 5K, there was no particular reason on Sunpower's part to expect that it would turn into the *Jarndyce* v. *Jarndyce* marathon that it did, nor was there any reason to expect that it would turn into a contract that would take Sunpower within a hair's breadth of bankruptcy. A million dollars in cash flow generally helps a company. But the horse-trading was brisk.

Sunpower's traditional bid for a project was straightforward: the project manager would calculate as nearly as one could forecast the actual costs of a project: how many hours of engineering, how many hours of drafting, how many hours of machine shop time, of technician time, of assembly and fabrication and testing time, how much for materials. This was toted up into a figure that the company reasonably expected it would be able to accomplish the job for. Then, as is standard practice in certain kinds of contract work, Sunpower would add a percentage to this figure for overhead, a percentage for "internal R & D" (that is, R & D necessary to do the R & D the job required), and a percentage for what Sunpower calls an "opportunity fee" and which Cummins would call, to William's vexation, "profit." This would be the total cost to the sponsor. In certain cases, there would also be a fee for a license, so that the sponsor would have the right to make and sell the fruits of the R & D—the engine, for example.

Cummins had, however, something slightly different in mind. First, there was the matter of the billing on the proof-of-concept engines, the two-kilowatt SHARP engines that Sunpower had refined and provided to Cummins for testing; that project was, from Cummins's point of view, over budget. From Sunpower's point of view, Cummins still owed Sunpower money, since Sunpower had spent more than Cummins had paid it to do the work. Then came the opportunity fee discussions, which at their height involved everyone who was anyone at either company in conference calls and faxes and meetings and letters and offers and counteroffers. Twice William went out on a limb and spoke directly with Jerome Davis himself, a chancy endeavor in that it might have been seen as undercutting the Cummins program managers and their authority. Cummins thought the opportunity fee too high and suggested a lower one; Sunpower gave a little, but insisted that at the Cummins figure Sunpower would lose money on the deal. The compromise figure would, Sunpower's John Crawford said, allow Sunpower to break even, something it was willing to do for a valued sponsor like Cummins. Cummins's Rocky Kubo insisted that Sunpower was going to make a profit. For a while, there was an impasse, an expensive impasse, because Sunpower had done many thousands of dollars worth of work on the SHARP/Cummins project and hadn't been paid for it. Further, Sunpower was paying the team members their salary, and paying the overhead, while the haggling went on, and there was no clear indication that they would be paid for this either, although Cummins was adamant that Sunpower needed to be working as fast as it could on the project instead of waiting around like day laborers to get paid. The percentages—they ranged between 6 and 15 percent in various permutations—were important to Sunpower because they represented the chance to actually end up after a year's worth of work with more money than it had started with, something that companies need to do if they are to survive; from Cummins's point of view, it wanted Sunpower's best efforts in terms of both activity and efficiency; if Sunpower did its job right, it would make a profit; if not, it wouldn't, which was Cummins's idea of what companies should do. The notion of an incentive arrangement thus crept into the discussions, and the impasse was solved.

At the time, this solution was thought of by many people both at Sunpower and at Cummins as a coup for their side, which should have been ominous in the extreme. John Crawford and Rocky Kubo worked out a complex list of incentive payments in lieu of most of the opportunity fee. Sunpower would bill on four quarterly purchase orders to cover time and overhead, but when the engine its team designed and built hit certain milestones, Cummins would pay a sort of bonus: ten thousand dollars for the first run of the first engine, ten thousand dollars for the first run at half power, ten thousand for the first run at the minimum overall efficiency, ten thousand for an hour run, and so on. Cummins liked this because it didn't mind paying for what it was contracting for: a working, debugged, well-tested piece of hardware. Sunpower liked this because it was bonus money in the truest sense— its people had no doubt whatsoever that they could hit the milestones easily and clean up with a bunch of cash. If everything went perfectly, Sunpower would stand to get big cash bonus checks simply for doing what it had said it could do.

The contract was signed and sealed, and almost the first thing that

Sunpower did was try the Box alternator, which absorbed a third of the money and a third of the time allotted to complete the task. In the massive binders Beej keeps of Cummins/Sunpower correspondence, the first thing following John Crawford's letter to Rocky affirming the details of the contract is a drawing of the Box alternator. Never have two pieces of paper been so wincingly difficult to look at side by side.

Chapter 10

Pig Dreams

Come on, Robi, that machine wasn't a pig.

—*David Berchowitz, in conversation*

Okay, okay, not a pig—a goat, then.

—*Robi Unger, same conversation*

THE 5K TEAM was now on the hot seat. Money spent on unworkable schemes was still money spent, and with the incentive plan, there was no cushion to fall back on. Make the engine work, or forget about the money.

The pressure was palpable; once, Lyn Bowman, flush from a particular success with one of the small Stirling coolers he had a hand in, slapped Neill on the back and asked him when the 5K team was going to begin pulling its own weight. Neill's reply was not something he would say in front of his mother. Neill's frustration was accentuated by the fact that they were making marvelous progress by most standards. The parts for the engine were in the shop, the assembly was progressing, there had been no show stoppers, no fundamental flaws. Later, Neill would find himself irritated by what he regarded as a lack of acknowledgment of what the team was doing and had been able to do; he theorized that because the project was so "fussy," everyone simply assumed that they would make it work, and thus expected much more than anyone ever expected from a first prototype.

The team was not being heroic in any familiar Sunpower sense—no all-nighters, no toggled-up stopgap fixes, no running on the edge of control. All were trying in the face of much adversity to be methodical, and that sometimes looked as though they weren't extending themselves very much.

So the first engine run of the 5K was not only anticlimactic but in some ways more frustrating than a failure would have been. The sodium-filled heat pipe, with its Sunpower-designed and -fabricated heater head already welded in, arrived and was tested. Dave Shade and Hans Zwahlen assembled the engine painstakingly and according to the scheme that Neill and Abhijit had envisioned when the parts were still all just on paper. First, the alternator was bolted to the test stand, the bearing surfaces inserted inside. Then the piston rod and the magnet paddle were lowered in, using one of the team's cleverest assembly jigs. This whole assembly was then carted into the test cell, hoisted up on a ceiling crane, and attached to the transition piece. Then the displacer and the displacer rod and the displacer gas spring. And the copper cooling ring and the steel "cooler shell"—the "cold end" of the engine. The regenerator and the displacer cylinder. Shade methodically checked each assembly, Neill over his shoulder; O-rings to seal the helium gas? piston and displacer rings for sealing the running surfaces? Everything was aligned and aligned again. Throughout all of this, onlookers were in and out of the test cell all day, keeping tabs, kibitzing, offering observations and witticisms.

The engine is hoisted up on cranes and lowered into the pressure vessel on end; then the cranes are shifted to lift the engine and pressure vessel up again as one large and heavy and unwieldy object. The engine hangs horizontally, parallel to the floor, so that six hundred pounds of steel can be attached at the back of the pressure vessel to absorb vibration. The engine is attached to the heat pipe, and dozens of hoses, wires, umbilicals, and monitoring lines are attached. Dave Shade purges the engine with a vacuum pump to get the air out, and fills the pressure vessel with helium. The heat pipe, powered by electricity, is turned on, and begins to warm.

Sunpower's oldest tradition has been the engine run, the first run, when everyone—everyone—gathers to see what will happen. Tradition dictates that the call is put out through the company, and everyone is present. But no call is issued at this moment; Neill sends someone to tell William that they're getting ready to run, but he doesn't assemble the multitudes. He expects that they will all trickle in at some point, but he has other things on his mind. Will it run? for example.

"It'll work," he says. "I don't know how it'll work, but it'll work."

The unanswered questions hanging in the air of the 5K test cell, as everyone stands around waiting for the heat pipe to warm up—it will take about an hour—fall into several categories. Some are endemic to R & D in general, others to a first-time event at Sunpower in particular, but all might be summed up like this: is each of the various critical paths properly converging?

Some of these critical paths are electronic, such as the monitoring and control systems for the engine and the heat pipe; others are dynamic and thermodynamic, having to do with how the engine as a physical and mechanical device matches up with its theoretical design. All of these paths, while tested to the extent possible without actually running the engine, are still untested under real conditions. Will the electronic "load controller" actually manage the power output of the alternator? Will the data acquisition system actually capture and reflect the data it was designed to acquire?

Still other critical paths are purely mechanical: inside the pressure vessel are many dozens of parts, many of which will shortly be subjected to loads and stresses that were calculated, but never tested in any conventional sense. Some of the mechanical parts are trivial to calculate for—the large springs inside the piston rod, for example, are selected from a chart of performance characteristics supplied by the spring company, and ordered through the mail.

Others are headache inducing in their complexity, principally the magnet paddle, that fragile ring of magnets wound in Kevlar thread and glued together with a frame of bicycle spokes. And there are crucial components, chief among them the electric "heat pipe" that is bolted to the engine's heater head, that were designed and fabricated with a minimum of input by Sunpower. A heat pipe is in essence an expensive, somewhat fragile, scary device that makes the heat that runs an engine in the test cell. Heat pipe design inhabits an entirely separate world from the world of Stirling engines, one that requires expertise and skills all its own.

The Thermacore Corporation, of Lancaster, Pennsylvania, is one of the nation's preeminent manufacturers of heat pipes, but only a couple of companies in the world are willing to design and fabricate a heat pipe such as might be used to test a Stirling engine. "Heat pipes," Dave Shade once said, "*never* work, and even if they *do* work, they *always* break." Neill Lane has gone so far as to suggest that Stirling engines are ultimately useful only as testing devices for heat pipes, because heat pipes attached to Stirling engines inevitably fail, often in interesting and unusual ways, and a Stirling engine program can thus generate all sorts of fascinating data about why heat pipes don't work. Or, when they do work, why they break. Always.

The Thermacore electric heat pipe that is now slowly warming up on the 5K is thus both babied and given a cold eye by all who are near it. A heat pipe is problematic because of the mechanics and physics of heat transfer: how to make something hot—evenly hot. Most of the heat pipe's bulk is made up of electric cartridge heaters, bigger and more powerful than, but not very different from, the heating elements on an electric stove. The cartridge heaters warm a sealed chamber that contains several pounds of sodium metal, a chamber sealed in vacuum—there is no air in the heat pipe, because sodium is so reactive it burns when exposed to air, and also to permit the sodium to vaporize at a lower temperature. The theory behind using a heat pipe is the same as that behind using a double boiler to melt chocolate: water is boiled to steam, and the steam condenses on the top pot of the double boiler, heating it in the process, and then drips back down as water into the pan.

In a heat pipe, the sodium melts, then vaporizes, and the vapor rises, heating the two hundred or so thin tubes in the heater head; the helium inside the engine flows through the tubes and picks up the heat that makes the engine go, and since the helium is taking heat out of the heat pipe, the sodium cools, condenses, and drips down to the bottom of the heat pipe, where it is vaporized again.

A heat pipe thus provides the best that engineering has to offer for transferring heat evenly from a heat source (the electric cartridge heaters in the test cell, or the concentrated rays of the sun) to a gas under pressure: the sodium vapor provides steady heat to the heater tubes, permitting even heat transfer and precise temperature control, but, equally important, a heat pipe prevents—again, theoretically—"hot spots," or places where the heater head or its tubes are subjected to wide variations in temperature across small distances. (This is that same reason why one doesn't melt chocolate by simply putting it in a pan on the stove.) Hot spots make heater heads fail; hot spots can melt the braze that holds the heater tubes in place, and thus provide a path for helium to leak from the engine into the heat pipe; hot spots stress bolts and welds and O-rings far beyond what they were designed to sustain; hot spots can destroy critical parts, or entire engines, or, indeed, entire engine programs (if the hardware burns up) in minutes or hours. Hot spots are bad news; heat pipes, theoretically, eliminate the potential for hot spots, because the sodium vapor puts the heat where the heat is being extracted—the heater tubes.

That's the theory. But it's easy to visualize what can go wrong. The heat pipe is a container of boiling metal at about 1,400 degrees Fahrenheit; a Stirling engine is a container of helium pressurized at about 600 pounds per square inch, with two hundred thin steel tubes protruding into the chamber where the sodium is boiling. The two containers are then bolted together, the mating point sealed by a pair of thin rubber O-ring gaskets and a ring of bolts. This may sound scary enough, but there's more. The heat exchangers of the engine are as close together as is possible, a function—a necessity—of design, so there is a temperature gradient ranging from something like 1,400 degrees Fahrenheit at the heater head, to something like 200 degrees at the cooler—just inches away. Ask a materials scientist what she might recommend as a way of creating a catastrophe, and chances are that what will come to mind is an arrangement very close to this: a very hot vessel at very low pressure attached to a comparatively cold vessel at very high pressure, with cyclic heating up and cooling down. It is right out of a textbook that might be called *What Makes Metal Deform, Weaken, and "Creep" toward Rupture*. This is not a book that anyone at Sunpower wants to write. Heat pipes respond terribly to the conditions they are most likely to encounter, principally leaks: helium in the heater head tubes at 40 bar (600 pounds per square inch) *wants* to find a way into the heat pipe, a place of subatmospheric pressure, but helium in a heat pipe can make a heat pipe explode, because the helium "tricks" the thermostatic controls into thinking that the engine is taking more heat out of the heat pipe than it really is. Beneath the heat pipe sits a disposable aluminum roasting pan filled with soda ash, the recommended substance to put on a sodium fire. "It's to catch the drips," says Shade, "after the sodium trickles down over the cartridge heaters and shorts them out. Only problem is that the drips will be dripping at 40 bar. Hey, Beej; what's the velocity of liquid sodium squirting out of a broken weld at 600 psi?"

Abhijit, studying the wall of gauges, starts to analyze the question before he recognizes that it is not a serious one; he shakes his head at Shade good-naturedly, and turns to the instruction manual that Thermacore thoughtfully provided with the heat pipe, a very thin manual, plumped up with photocopies of the wiring schematics. "Instructions for bench top operation of sodium heat pipe," Shade reads aloud over Beej's shoulder. "Not on *my* bench top."

Hans Zwahlen chimes in: "One: turn heat pipe on. Two: run like hell."

"You guys . . . ," Beej says, but Hans begins telling stories of heat pipes he has known, most of which burst, burned, ignored all attempts to modulate their temperature, ran too hot, too cold, not at all; the stories generally end with Hans describing, as graphically as he can for Neill's and Beej's benefit, the sound and sight of a sodium fire. "*Biiiig* clouds of smoke, man; it's just *amazing*. You should see it." Shade takes up the cause. "Don't worry; you will." Neill goes out into the hall to have another cigarette.

Electronic controls. The heat pipe. What else at a first engine run? Well, the engine might not run, for one, or dozens, or literally hundreds of reasons. Or the engine might run—*away,* overpower the alternator, smash itself to bits. It might *try* to run, overcoming some unforeseen or unimagined mechanical interference or binding, and in doing so, ruin the magnet paddle, or the displacer, or the piston rod, or. . . . Or it might run and then quit because of a failure of some part that couldn't take its theorized load, or that was fabricated out of the wrong material or simply a bad piece of the right material, or that was unknowingly weakened or damaged in fabrication or assembly or. . . .

Or it might run, in descending order of disgrace, like a dog, or a goat, or, like a pig, the lowest creature in the Sunpower animal kingdom.

A pig of a machine is a painfully immediate metaphor, an engine that huffs and puffs, piglike, chugging along at a quarter or a third of its design output, no matter what anyone does to it to make it better. (The pig-*shaped* pressure vessels in which Stirling engines reside must contribute to this metaphor: the SHARP engine on display in the Sunpower lobby, with its flat heater heat of a snout and beady control feed-through eyes, even has a curled thermocouple wire for a tail. It looks as if it were about to roll in the mud.)

It takes extreme life-giving and life-support measures simply to get a pig to start, or stop, or run without breaking its weakest link. A pig uses huge amounts of heat to make very little electricity. A pig is wincingly inefficient, noisy, arrhythmic, hot to the touch, too scary to be around when it's running, and too depressing to be near when it isn't. A pig sends engineers to the West End Tavern so that they might drink to forget it.

Engineers aspire to high art, but engines, inventions, devices that push the envelope, can still be, as a result of fate or misfortune, lousy pieces of hardware. Only God can make a Stirling engine, but it takes a design engineer—and a million dollars of someone else's money—to make a pig. Of all the dreams dreamed by Sunpower engineers, only Lyn Bowman and David Berchowitz admit to having dreams of Stirling machines running perfectly; everyone else wakes up thinking, *"Oink, oink, oink."*

Neill stubs out his cigarette and reenters the test cell. "Let's motor it," he says to Shade. Motoring an engine is a last-step-before-running test procedure, in which current is fed into the alternator to make it oscillate; this causes the alternator to move the magnet paddle, which moves the piston, which compresses the helium, which bounces the displacer, and so on: motoring a Stirling engine is exactly the way Stirling-cycle coolers work: electricity makes the linear alternator oscillate, and the Stirling cycle of expansion and compression of helium by the piston and displacer happens, only instead of turning heat into electricity, a Stirling cooler (or a Stirling engine being motored) turns electricity into the "reverse" of heat, which laypersons call "cold." An engine is motored before running for several reasons, but primarily to make sure that all its parts are moving freely. It's sort of like pushing an automobile a few feet before trying to start it.

The motoring test is fine; on the oscilloscope screen, a green oval appears, tilted over on its side. It describes the positions of the piston and displacer in space, and relative to each other—a "Lissajous" plot, a graphical method of representing the phase relationship of piston and displacer. The trace looks like an oval because the piston and displacer complete one full cycle every sixtieth of a second; if the engine was motored at, say, one cycle per second, the traces would become a dot moving lazily in an oval. The value of such a picture is incalculable, the equivalent of seeing inside a device while it is working; it is accomplished through a clever invention by Robert Redlich, one of William's oldest friends and one of Sunpower's most treasured consultants. On the piston and the displacer assemblies are devices called FLDTs, or fast linear displacement transducers.

An FLDT looks like a small aluminum rod in a small aluminum sleeve; move the rod, and the electronics of the device measure the "displacement," or movement. Hook the FLDTs up to an oscilloscope, and "watch" the piston and displacer move, in real time and in actual phase relationship to each other. "A picture?" Neill asks, and Hans Zwahlen gets out a Polaroid camera with a special shroud for taking photographs of Lissajous plots, holds it to the oscilloscope, and pulls the camera trigger. Even as the film develops, Neill, Abhijit, and Dave Shade all stare at the oscilloscope and the computer screen above it. What they are seeing gives them good hope, good cheer: a robust loop on the oscilloscope, a respectable phase angle between the piston and the displacer, respectable amplitude on piston and displacer, modest power consumption, which means modest friction between the parts inside the pressure vessel that are rubbing together.

They turn the power off, and on again, watching the Lissajous plot to see if the alternator, or the bearing surfaces, or the ring surfaces, seem to be sticking or grabbing. Everything looks fine, but for a tendency for the piston to "run in," or find its preferred centering point somewhat closer to the "in," or hot, end of the engine than to the center of its cylinder. Shade remarks on this in a neutral voice, since he has been raising this possibility all through the fabrication process.

At lighter moments, he has been lobbying Neill about the need for a "centering scheme" (since David Berchowitz has over time perfected several of the most common methods used at Sunpower to center pistons, the methods are invariably referred to as "schemes," a word that rolls off a South African–educated engineer's tongue), since Dave Shade has never seen a Stirling engine that didn't need some strategy— other than pure mathematics—to have the piston center itself in its cylinder. "We'll just have to see" is all Neill says at this moment, although now that they are all watching it so closely, it does indeed seem that the piston "wants" to run in. "Centerports," Shade says, to which Beej, ever the conciliator, says simply that they'll cross that bridge when they come to it, meaning, first, that engines don't invariably run the same way that they motor and, second, that, well, the engine hasn't even *run* as an engine yet. There might not be such a bridge to cross.

As the heat pipe heats, they run through the various procedures and fall-back procedures for this first test. As they do so, there is a steady stream of people in and out of the test cell; William, Joerg, Barbara Fleck, Jarlath, David Berchowitz. Neill sets what everyone thinks are very conservative limits for the first run: 35 bar, or about 500 pounds per square inch on the pressure, and 600 degrees Celsius (1,100 degrees Fahrenheit). First runs always start out conservatively; sometimes they even finish that way.

With all the hardware assembled, there are only a few things that can be done to affect the performance of the engine, but they are an important few. One is the ability, via the electronic load controller, to vary the stroke of the engine: longer stroke, more power. But stroke is limited by the physical spaces within the cylinders—too long a stroke, and the displacer will hit at the end of its cylinder, and the piston will hit the displacer. The resonance and motoring tests suggest that the piston is running in slightly, which has the practical effect of shortening the maximum stroke at which the engine can run. These are mechanical limits; when the displacer hits the end of its cylinder, or the piston and displacer collide, everyone in the test cell knows it, can hear it, and Beej will back off on the stroke until the banging stops.

The other two important controls are heat and pressure. More pressure means, in a closed system, adding more helium; one of the effects of this is to increase the frequency at which the engine operates (and thus get more cycles of work out of the machine); another effect is a function of the "Beale number," or the relationship between the volume of the work space in front of the piston and the mean pressure of the machine; a third effect is simply the transfer of more heat by putting more helium in to carry it. Raising the temperature of the heat pipe is a way of directly affecting the ultimate efficiency of a Stirling engine: the higher the temperature, the more power a machine will produce.

Up to a point, that is, which is why Neill sets rather conservative limits to start. Both heat and pressure are limited by the mechanical integrity of the engine and its parts—the heater head has a finite mechanical strength, an upper limit to the temperature at which one can be certain it will not fail. Also, the mechanical integrity of the heat exchangers limits the pressure at which an engine can safely be operated; the cooler in the 5K is the weakest link in terms of mechanical strength, fabricated as it is out of copper and sealed with rubber O-rings. There is a design temperature and a design pressure for any machine, a specification that all the parts have been designed to meet. The design temperature and pressure can be exceeded, of course, without an immediate catastrophic failure, but no one can say exactly what the failure temperature or pressure would be.

On the other hand, it takes an engineer with large amounts of decidedly un-Sunpower-like discipline not to want to squeeze a bit more power out of the engine; paradoxically, an engine running near its design output without being squeezed is not nearly as tempting to squeeze as one running at, say, half or two-thirds of design output. This is when it becomes tempting to raise the temperature a little, raise the pressure a little, a little more heat, just a little more pressure . . . and, first thing one knows, the engine is running at some gray area of temperature and pressure that causes people who have had strength-of-materials classes to do shirt-cuff calculations. It also can have the practical effect of making people back away from an engine. Once, when William was running a small-engine prototype with Al Schubert as his technician, William took control of the helium valve himself, coaxing, coaxing, coaxing, a few more watts out of the engine by increasing the pressure, but he was still nowhere near where the engine should be. He looked at the round pressure gauge on the engine, where the needle was down near the zero, and said to Al in disgust, "There's *no* damn pressure in the machine *at all*." Al looked at the gauge and saw that the needle was *all the way around the dial,* stopped only by the pin at which it normally rests, but on the back side, as it were. Al suddenly decided that he needed a breath of air, and he wanted to take it as far away as was physically possible.

When someone as sensitive to the practical limits of hardware as Al Schubert leaves a test cell, it takes a certain amount of nerve to stay inside. (Or a certain naïveté; there have been engine runs for prospective sponsors where it seemed that everyone at Sunpower was *extremely* courteous, leaving the sponsor alone in a test cell.) William, though, is absolutely fearless, having seen more hardware fail than most engineers ever see running. Just as he has an intuitive sense for how and why something might work, he also has an intuitive sense for the limits to which hardware can be pushed, and that he is still alive and fearless is testimony to this. He is a hardware man, a toucher and tweaker of machines. Once, when the 5K heat pipe had just been tested by heating it up to where it glowed red, William ambled into the test cell to join the group admiring it. "It's still hot," he was warned, which prompted him to touch the top of the heat pipe for a moment with his index finger. "Humans have not evolved yet to the point where they can resist touching something that they've just been told is hot," he said matter-of-factly, touching it one more time and then ambling away. The heat pipe at that moment was at 1,200 degrees Fahrenheit.

A man who is on such intimate terms with heat and pressure will use them to find the limits of a machine. This, however, is something that Neill emphatically doesn't want to do, not this day, not at the first run. Reasonable temperature, reasonable pressure, see the power, see the efficiency, get good data, and then take the machine apart and see what it looks like. It is not putting too fine a point on it to say that if there is a catastrophic failure at this stage—if the magnet paddle disintegrates or the alternator explodes or the displacer jams in the heater head and welds itself there—the 5K team will be out in the street. It needs not great news but good news. And good news will consist of an engine run at *any* power: that's all that's necessary for the first incentive payment; if things look fantastic, they can consider running for an hour, which will be the second incentive. But no heroics, not today. Abhijit concurs; he doesn't want to make a phone call to Cummins in which he says, "We hit the first milestone just before it blew up."

The heat pipe is nearly to operating temperature when Eric Bakeman and Morgan Mitchell come in to look at the data acquisition system, the computer-based monitoring equipment that will store all of the relevant data from an engine run and that he and Morgan have designed and—mostly—debugged. They are helped by Shade, who points out, from his long experience at watching computer screens flashing engine data, which numbers look "real" and which look "flaky." "The thermocouple there with the *two* decimal points looks flaky," Shade says, deadpan. Indeed, one of the temperatures is being recorded as "26.5.3," an unconventionally precise notation. "Just ig-

nore the second decimal point," suggests Eric. Among computer programmers, this is known as a "quick fix"—a "short-term fix" would be
covering the second decimal with a piece of tape.

Neill and Abhijit are more concerned with the reading for the
phase angle between the piston and the displacer. Recorded in degrees,
the phase angle has, when the engine is motored, a minus sign in front
of it, suggesting that the piston is leading the displacer, rather than vice
versa: is the minus sign real or flaky? Eric punches buttons, and the data
acquisition screen disappears, replaced by the programming code for
the software, an event that disconcerts Abhijit, who has been watching
very carefully the temperature on the heat pipe. "How long are you
going to be, Eric?" Abhijit asks, fully expecting that at the exact moment Eric took away Beej's ability to monitor the heat pipe, the heat
pipe decided to go berserk. Hans has already discovered with great glee
that the control dial on the heat pipe works only when it is jiggled, a
piece of knowledge that makes everyone even more delighted to be
around a heat pipe again. Hans pretends to read from the heat pipe
instruction manual. "Jiggle once for on, and twice for off," which gets
him a dirty look from Abhijit. Eric restores the data screen and says, "I
can't fix it if I can't *see* it; it's probably flaky anyway."

"Probably," says Neill, "*how* probably?"

Morgan Mitchell says emphatically, "It's *definitely* a flake." He
pauses. "I think." He and Eric then have a spirited discussion in the
secret language of computer programmers, full of talk of loops and
multiplex and terminating strings. They conclude by saying that most
probably it is a flake. They think. The round-robin discussion begins
again. "A flake," Eric says with finality. "I'm 90 percent sure its a flake."

"It can't be real," says Shade. A pause. "Could it?"

"No," says Abhijit. A pause. "I don't *think* so."

"It's *got* to be a flake," says Morgan. "The engine can't *run* with a
negative phase angle, can it?"

"It's wrong," says Eric, with absolute certainty, to which Beej asks,
"But *what's* wrong? The phase angle or the program?"

"Well, the *sign's* wrong," says Eric. We know that. . . ."

"Forget it," says Neill, "Just forget it for now. It can't be real. Just,
can you fix it when we're done here?"

"Sure," says Eric, affably. "When are you going to run the engine
again?"

"They haven't even run it *once*," points out Hans, which this time
earns him a sardonic look from Abhijit.

"Thanks for that vote of confidence, Hans."

"Oh really?" says Eric. "I thought you ran it a couple of weeks ago; weren't you supposed to?"

This sends Neill out into the hall for another cigarette, after first determining if Eric is kidding or not (he is) and then telling Eric what he should go and do to himself. But just then William saunters up and says, "Well?" and he and Neill go into the test cell to see what will happen next.

Chapter II

Data Acquisition on a Non-Ultimate Machine

The efficiency of the machine based on electrical input to the heat pipe and electrical output from the alternator (a very conservative measuring stick) is 26 percent.

—*Neill Lane, reporting on an early 5K engine run at the Monday morning meeting*

That's one-quarter as efficient as a piece of wire.

—*Dave Shade's interpretation*

T HE 5K STARTS so quickly and simply that virtually everyone in the room is startled. "Wow," says Eric Bakeman. "Didja know it was gonna be this easy?" They had run through their strategy one last time, positioned themselves at their various stations—Beej at the load controller, Shade at the load bank of electric light bulbs and the helium valve, Neill in front of the data acquisition screen—and after they all shrugged, Beej flipped a switch that sent a jolt of electricity into the alternator, and the engine started right up and ran. QED. The tension in the room vanishes for about thirty or forty seconds while all marvel—trying not to look *too* amazed—at their good fortune, at their abilities, at the Lissajous plot on the oscilloscope, at the numbers flickering on the data screen, at the fact that after all the work, after the efforts of a half dozen people over the course of a year, what happened was exactly what should have happened. Flick a switch, and the engine comes on. "We've got to quit acting so surprised," Beej says, but he is smiling. "What clever chaps we are," replies Neill, also smiling. There is some handshaking and congratulating of one another for a few moments, and then they all arrange themselves around the oscilloscope and the data screen and start looking a bit more closely at what they have done.

What they have made is an engine with a gross design power output of something on the order of six kilowatts, which should produce something on the order of five kilowatts, but which in its first moments of life is producing about two kilowatts—although it is not running at its full stroke. They wanted to start the machine off slow, gently, and they have done so. As it warms up—that is, as the helium inside warms up and begins to expand—Dave Shade lets some helium out, and Beej adjusts the stroke upward, eight millimeters, then nine, then ten. At ten millimeters, the power out is about two and a half kilowatts.

Beej strokes it out a little more, up to twelve millimeters amplitude, which gains several hundred more watts. The 5K team is making electricity. All can see they are making electricity, because on the wall of the test cell a sheet of plywood covered with a half dozen rows of light bulbs and two rows of electric stove burners is giving off lots of light and a fair amount of heat—they are making electricity, but of course they aren't selling it. They are consuming it, dumping it, wasting it, in the test cell, the light bulbs and stove burners serving as the load on the alternator.

At this moment they are not producing a *net* gain in electricity either. The heat pipe is consuming something on the order of twenty-five kilowatts as it supplies heat to the engine: twenty-five kilowatts into the heatpipe, some fraction of that into the heater head, a fraction of that into the helium, a fraction of that to the piston, a fraction of that to the alternator, and what is left to the rows of light bulbs. Power Out divided by Power In is "efficiency," or how large a fraction of energy a device can convert into work. As the engine warms up further still, and Beej strokes it out a little more, Power Out creeps up, up, up, until Beej can increase the stroke no more without having the piston bang into the displacer.

When he tries, at a bit past thirteen and a half millimeters, there is an immediate clanging from the pressure vessel, and he backs the stroke off immediately. Thirteen and a half millimeters, and about three kilowatts. "What's the frequency?" someone asks. (When Dan Rather was mugged several years ago by a man who asked him repeatedly, "What's the frequency, Kevin?" everyone at Sunpower claimed that the mugger was a Stirling-machine designer gone berserk. It is a commonplace question in a test cell.) The frequency is a bit low—around 55 Hz, or fifty-five cycles per second. "More gas, Dave?" asks Neill, and as Hans and Shade both chant, "More pressure, more pressure!" Shade opens the valve, and the frequency and the power edge up. "More heat?" asks Shade, but Neill shakes his head; not yet. Maybe not at all; he is trying to get a sense, a feel, for the machine. Three kilowatts and no major problems is great for a first run, but it is not five kilowatts. The engine has run for about ten minutes when William asks, "So the power is down, hm?"

Neill is noncommittal; they are still feeling their way. Shade notes that the piston is indeed running in, about one and a half millimeters. "Take it apart and put in centerports?" he asks, but again Neill is non-committal. "A picture, Hans?" he asks, and again Hans covers the oscilloscope with the camera shroud and snaps the trigger. The phase angle—the relationship in position between the piston and the dis-placer—is a little low, but not worryingly so. "Stroke it out?" says Neill, and Beej edges the piston another half millimeter in amplitude; no hit, but at least from the plot on the oscilloscope, the piston and displacer must be within tenths of millimeters of each other. Thirty-one hundred watts, and Beej mentions conversationally that the "back end" of the machine—the alternator—is warm, not hot, but warm.

Neill Lane stares at the data screen. "The power is there some-where," he says, half to himself, and Shade agrees. "If you could stroke it out, you'd be close." He pauses. "Well, closer."

"You have a *negative* phase angle?" asks William; he has noticed the minus sign on the data screen; ten people answer his question with varying degrees of specificity. "Go up in temperature to six twenty-five?" asks Neill, and Beej adjusts the heat pipe control, jiggling it furtively as he does so. In a few minutes, the heat pipe is up to 625 degrees Celsius, then 627; another eighty watts. Beej again mentions the back-end temperatures, which have crept up a few more degrees; nothing serious, but worth noting.

An alternator in a Stirling machine can grow hot for a variety of reasons; it will always grow hotter than ambient temperature, since making electricity is never perfect and imperfect processes generate heat as a by-product, but there are more ominous things to read into a hot alternator. Most ominous would be a rub between the magnet paddle and the laminations, causing friction. A rub can be the result of simply a lack of concentricity—imperfect cylindrical shapes, rather than the theoretical ideal—but also a result of "side loads," which occur when the magnet paddle, rather than riding exactly in the center of the narrow gap between inner and outer laminations, rides closer in one place than another, close enough for the magnets to be drawn to the iron laminations.

In theory, an axisymmetric linear alternator has no side loads. If all the pieces were perfectly round, this would be true. But nothing in the alternator is perfectly round—indeed, the magnet paddle, when it is inserted into the air gap during assembly, must be carefully placed in a particular orientation, or there is a rub. The team members knew this from the moment they first assembled the alternator, knew they didn't have perfect circles. But there is an orientation where the magnet paddle seems not to rub, and this is the orientation in which it is installed. But whether the paddle is free or rubbing while the engine is running is something that can't be known while the engine is running except by inference, inference based on information such as the temperature of the laminations, which creeps up another degree.

"Let's turn it off," says Neill. There doesn't seem to be more power readily available, and they might be wearing a bad patch into the magnet paddle to boot. "Should we use the *kill valve?*" asks Shade, to which Beej says, "I was just going to suggest that."

The kill valve is a large red button on the control panel; pressing the button should open up a valve in the displacer gas spring, and without the displacer gas spring, the engine shouldn't run. Beej presses the button, and the Lissajous plot shudders and then shrinks to a dot. Beej beams. The kill valve works.

Then they are shutting the heat pipe down, conferring about the data, suggesting possibilities. Peggy Shank comes in and asks, "*Party?*"—Peggy being the official party organizer at Sunpower. But there is an odd lack of elation in the room; perhaps if they weren't so far behind schedule, perhaps if they had had a *big* win—five thousand watts—perhaps if they had not proceeded so steadily and in such an organized fashion, if instead they had needed some heroics to make the engine run, they might feel different.

But they are in a limbo very familiar to everyone in the room: the engine runs with nothing obviously wrong, but not the way it should. They know exactly what to do, but what they will have to do is a tedious and iterative process made more tedious and plodding by the sheer logistics of disassembling and assembling the engine. Cool it off (overnight), take it apart (which takes most of the morning), and look at it after lunch. A quick change can mean getting the engine assembled again for a run the next morning, but anything even slightly tricky can mean two and three days between runs.

This is why it takes discipline to not change three or four things at once; if a half dozen changes are made and the engine runs brilliantly, they still wouldn't know what they did that was right, wrong, or neutral. Change six things and turn the engine into a *real* pig, and the only sensible thing to do is go through them one by one, and that means wasting the time ostensibly "saved" in the first place. They have also, without destroying the hardware or digging themselves into a hole, hit the first milestone in the Cummins program, and have earned a modest incentive check, just by running the engine. Three kilowatts, no show-stoppers, no disasters.

The question now is what to do next, or what to do first. "Take it apart, and we'll look at the magnet paddle and the bearings," says Neill. "And the rings; let's start out by seeing what shape it's in, and we'll go from there."

Shade, though, has a suggestion. "Listen for just a minute; you're gonna need centerports anyway, right? I was looking at the drawing. Here." And Shade sketches out on a scrap of paper a centering scheme, very similar to one that he implemented on a previous engine at Sunpower. "A collar here, and it's easy to clamp with a ring, like this. Two holes, and then you've got this . . . sleeve on the piston anyway, so you Xylan it. It's simple parts, it doesn't ruin anything; four hours in the shop and it'll work."

Neill is wary; centerports, like other things in a Stirling machine, are fine in theory—a controlled leak between work space and bounce space, a way of intermittently letting just enough pressure through to center the piston. Centerports are in and of themselves self-centering, in that when the piston is running in its proper place, they don't allow any gas to leak at all; only when the piston runs in do the centerports open and bring the piston to where it should be. But the placement of the centerports is more guesswork than anyone likes: in essence, the holes are drilled on the basis of the distance that the FLDTs and the data system say the piston is running off center, distances measured in tenths of millimeters. Centerports that open too soon keep the piston running out; ports that open too late don't do the job. And no matter how carefully one measures, the first try is just that, a first try. Guesswork, which they have been trying to avoid.

But Shade is persuasive; his scheme is simple, straightforward, and easy to implement. Neill nods. "All right, tear it down, we'll look it over, and we'll put in centerports."

Shade grins. "Told ya." He enters the test cell to take the glass wool insulation off the heat pipe, and arranges with Hans to meet him at eight sharp the next morning to take the engine apart.

Over the next six weeks, Shade and Hans take the engine apart and put it together enough times for it to be a sort of routine, or ritual, as the team looks for the errant kilowatts.

The centerports worked the first time they were tried; Shade's clever design, while taking several days longer than his predicted four hours to implement, did indeed cause the piston to run more or less at center when they next ran the engine. But the power was, really, no better: the engine ran fine, but was underpowered. Why?

The likely place to look, after a series of runs to implement the centerports, was the bearing sleeves on which the piston and displacer rods moved. The bearing surfaces they were using were still bronze coated with Xylan, originally intended just for the resonance test, but the real bearing sleeves hadn't arrived from the company that was making them—Neill had visions of starting up a conversation with someone in the unemployment line with him who says that he worked for a bearing company that went bankrupt because it kept losing customers' orders. The bearing sleeves they awaited anxiously were extraordinarily expensive and painstakingly custom-made from an exotic carbon fiber—poly-ether-ether-ketone, or PEET—two entire sets of bearings made from slightly different PEET recipes. The bearings took on an almost mythic significance the longer their arrival was delayed. The power *must* be in the bearings. The bearings used in the first runs didn't look bad—they were "bartched," or rubbed shiny—in places, but nothing awful; still, the suspicion was that they were costing a lot of power through friction.

Joerg Seume and Hans and Abhijit spent two days with the magnet paddle and alternator assembly in the assembly cell, doing tests on the old bearings, trying to measure the friction between the bearing surfaces. On the third day of their tests, it so happened that the PEET bearings arrived, in the two different versions as specified; one set was a slightly different formula of PEET, and a lot more expensive, but supposedly better. The bearing sleeves were of an unusual design and shape as far as the PEET people were concerned, which is why it took *seven months* for them to be delivered, six months longer than the contractor had said. Abhijit and Joerg tore open the boxes and boxes within boxes with the enthusiasm one might expect after waiting half a year. Joerg took one of the allegedly magical sets of bearing sleeves out of its little jewel-box packing case and, while telling some story to Abhijit and Neill, locked the jaws of his vernier calipers across it, a casual checking of the dimension. He stopped telling his story and looked down at his notebook where he had recorded the dimensions of the bearing sleeves as ordered. Then he measured the sleeve again. Then he rustled through his notebook to find the photocopy of the original order sheet, and looked at those dimensions. Then he measured the sleeve again. By now he had drawn a crowd. "Dave," he said to Shade, "would you measure this please?" Neill knew this was not going to be anything like good news, and he turned his face to the wall and began cursing. Shade grabbed his own calipers and measured the bearing sleeve. "Eighty-four millimeters on the OD," or outside dimension, he said. "What did you get?"

Joerg taps his calipers against his upper lip. "That's what I got, too." He reaches into the packing case and retrieves another magical sleeve, measures it, and taps his upper lip again. "Eighty-four. Well, gentlemen . . . ," says Joerg, but Abhijit already has a ruler against the 5K drawing on the cell wall. "Looks like eighty from here," he says. Seven months is a long time to wait for something that is useless.

There is still the other set of PEET bearings, the plain vanilla PEET, to check. Joerg unpacks these and measures; they look to be the right size, so it's not a total loss, but instead of two options, they now have one. Joerg suggests he go call the PEET man. "Ask him if he's a bloody moron," Neill suggests unhelpfully. "Ask him if there are any jobs there for engineers who can read. Ask him if . . . ." But Joerg doesn't ask the man anything; he is out to lunch. "I hope he *chokes*," says Neill. "Are the other ones all right?"

The regular PEET sleeves are exactly to spec, and Hans and Shade set about tearing the alternator and rod assembly down so that they can install them. Hans mentions that the new bearing sleeves, while breathtakingly expensive and supposedly good for forty thousand hours, "look like little rickety pieces of shit," and Shade concurs. The sleeves do seem delicate, because they are so light; they look like metal, but feel like *plastic*. When Shade snaps in the retaining rings that hold the sleeves in place, he does so as though he were performing delicate surgery; still, when the ring snaps around the bearing with a loud *click*, everyone winces. Since, fortuitously, they had been testing bearings for friction when the new bearings arrived anyway, they rig the alternator assembly up to test the PEET bearings; if they seem to cause much less friction in the same test than the previous bearings did, the 5K team may have found some watts.

The test of the bearings is a harrowing-looking procedure. The alternator and magnet paddle are bolted to a test stand in the cell, and a Variac is hooked up to the alternator leads. When current is introduced, just as when the engine is motored, the magnet paddle oscillates. Small amounts of current make for small oscillations—two millimeters or so—but since the alternator is drawing a dozen amps of current, not much more can be added. While the Variac is on, the alternator assembly, in places, is "live"—touch it while grounded and get a dramatic shock. So no one touches it. An ammeter is used to measure the amount of current it takes to make the magnet paddle move, an indirect measurement of how much friction there is in the bearings. The current is turned on and off, readings are taken, discussions ensue. And then there is a curious sound, a crackling, like an electric short. Shade kills the current immediately, and everyone looks at the alternator. "What was it?" asks Neill, when Shade can find no short. Shade shrugs. "I dunno; somethin' in the Variac, maybe?"

They turn the current on again, and again there is a crackling. Current off. Careful examination. Nothing amiss. Neill leans his ear right over the alternator, and tells Shade to turn the current on. This time the cracking is louder, and Neill looks down into the top of the alternator, down at the new PEET bearings. Abhijit leans over and looks as well. As they watch, what they see is amazing, so amazing that they both watch for several seconds before both shouting at Shade to turn the current off. Beej reaches into the collar at the top of the bearing sleeve and with a fingertip, takes up a tiny chip of poly-ether-ether-ketone bearing sleeve. The forty-thousand-hour bearings that took seven months to arrive have disintegrated ten minutes after being installed. "JOERG!" Neill calls, and Joerg comes into the assembly cell. "What's up?"

Neill shakes his head and takes out a cigarette. "Ehhh, don't worry about calling the bearings guy," he says, and he goes out into the machine shop to smoke.

Why the PEET bearings disintegrated so easily was never nailed down for certain—there was no time, really, to consider it. Most likely, the peculiar forces of being clamped in place and then subjected to the force of the piston rod moving linearly subjected the bearings to a load that was unpredictable in their design or fabrication. Whatever the reason, they were useless; for a couple of hours, the team experimented with the idea of gluing the bearing pieces together, but Beej took one look at the bearing sleeves covered with glue and said, "No way." It would be foolish to put something like that inside the machine, foolish and potentially catastrophic.

After Hans and Shade tore the alternator assembly down for a third time that day and reinstalled the bronze and Xylan bearings, the magic bearing sleeves were stacked on a shelf in the assembly cell and never thought about again. Nearly a year later, Larry Nelson, the handyman at Sunpower, threw them out in a fit of tidying up, when no one he asked could think of a reason for saving them: custom bearings come with no warranty.

With the bronze and Xylan bearings, team members feared they were perhaps seeing a mechanical limit to the power of the 5K machine. The fat, hollow, hardened steel piston rod, clever as it was, was heavy, there was no doubt about this. At the time, no one could conceive of how else to rig the bearings, so the hunt for kilowatts turned elsewhere. Shade re-Xylaned the bearing surfaces and reinstalled them, and the engine entered its "run it, tear it down, change something, run it" phase again. Next, they put centerports on the displacer—the piston centerports had caused enough of a change in the phase angle that the displacer was affected. Then the steel springs on the displacer rod were eliminated in favor of a larger gas spring, in an effort to lighten the load on the piston and, hence, on the piston bearings.

At each run, there were incremental improvements in power, but there was no big change; calculations and analysis of the data showed that the power was going in and that the alternator was turning proper amounts of that power into electricity. The power was there, but they couldn't find it. Tear it apart, put it back together, run it; someone stuck a cartoon on a bulletin board that showed an auto mechanic standing in front of a large furry animal smiling from the engine compartment of a car. The mechanic is saying to the owner, "*There's* your problem."

Then, in the middle of this exacting and time-consuming and frustrating process, they dropped the engine—all million dollars and four hundred pounds of it—onto the concrete floor of the test cell, late on a day that wasn't very good to begin with.

When they figured out how it had happened, the most puzzling thing about it was why it hadn't happened before. Every time the engine was assembled, it was trundled into the test cell on a wheeled cart and raised into the air with a chain hoist, by hand. The big hook of the chain hoist was clipped into a steel eye called a clevis, which itself was attached to the top of the engine assembly with steel cable run through a pair of eyebolts. All of this was very massive-looking equipment, the sort of cables and hooks that would (literally) hold a battleship at a dock. The chain on the hoist would be pulled, hand over hand, and the block-and-tackle arrangement of the hoist would propel the engine up, until it was hanging vertically. Then it was lowered, hand under hand, into the pressure vessel, and the transition piece—on its own chain hoist—was lowered down, and bolted into place. More chain hoists were attached to the pressure vessel itself, and the whole arrangement hoisted up yet again, until it was off the floor and could be, through a raising and lowering of the various chain hoists, suspended horizontally in the test cell.

Everyone even remotely associated with the 5K program had, in the course of two dozen engine runs and disassemblies, taken a turn or nine at running the engine up the chain hoists, then running the engine down the chain hoists in the reverse of the procedure.

It was during the very last part of the disassembly procedure that the engine and alternator assembly crashed to the floor of the test cell with a thud that could be heard—and felt—in the most remote offices of Sunpower, Inc. What had happened was this: when the engine and alternator assembly was being lowered onto its wheeled cart, there was always an odd and frankly frightening moment when it had to be tipped onto its side while still hanging from the hoist—it had to go from hanging vertically to lying on the cart horizontally. The queasy moment occurred when the engine was resting on the cart and its mass shifted from the vertical to the horizontal, as if one were tipping over a garbage can. The chain hoist and cable would jerk, but it would never—could never—be stressed to the breaking point. So the engine would tilt, and then tip, and then be lowered down. But the massive hook on the chain hoist—rated at two and a half *tons*—has a slender strip of metal that closes over the open part of the hook, a piece of metal rated at something like sixteen ounces; it is there merely to keep the cable from slipping out when the cable is loose, but it is not meant to hold any weight. On this afternoon, when the engine was being lowered down, the hook was turned backwards, so the cable, and the mass of the engine, was resting on the little strip of metal, rather than on the massive hook.

When the mass shifted, the bottom of the engine skidded out when the little strip of metal gave way, and the engine took a nosedive, crashing to the floor from a height of about four feet with the solid sound of finality. Abhijit, in the control room of the test cell with his back turned, thought that the steel I beam in the ceiling had come crashing down, so impressive was the shudder of the floor beneath his feet and the clamor of metal hitting concrete. Neill, out in the hall smoking a cigarette, expected that some technician's wife was now a widow, so frightening was the shudder of the floor beneath *his* feet. In the hallway outside the conference room—a hundred feet and three sets of doors away—two engineers looked at each other and said, "Was that thunder?" In the cooler test cell at the extreme far end of the building, a technician was startled enough to burn his thumb with a soldering iron.

At moments like this, it is only humane to report that the first words out of Abhijit's and Neill's mouths when they ran into the test cell an instant after the crash were unequivocally "Is anyone hurt?"

No one was hurt, just very surprised—since it wasn't immediately apparent *why* the engine had fallen, it had seemed for a few seconds like some act of God. Surprised, and sheepish.

The next thoughts that occurred to Abhijit and Neill were perfectly understandable, although both men had the graciousness to ask, twice more, whether anyone was hurt in any way; what they were both thinking was, "How is the engine?" Dropping it nose down onto concrete was not in the design specification; no calculations had been done. To have babied all those parts, dreamed of them, carried them carefully from room to room, wiped them down with clean cloths and alcohol, for week after week after week, it was distasteful to think of it all ending this way. "Ehh," said Neill, "I um—are you sure everyone is all right?— um, how do you suppose the engine is?" They began to look, the way one peeks through fingers at a horror movie.

Neill and Abhijit had always despised the tip-over-the-garbage-can method of disassembling the engine and had several times imagined what it would be like to watch a quarter ton of metal fall through space; they had all manner of jigs and rigs and procedures and safety strategies for literally every other step in the assembly process, but they had never been able to do much with that moment when the mass of the engine is shifted from vertical to horizontal; the most they had done was overspecify the chain hoists and cables and cleves so that nothing would collapse, but the weak link turned out not to be in the chains at all. It was a little strip of metal that Shade, once he looked at it closely, was able to bend back and forth with his fingers.

"Someone go and tell William that we dropped the engine but nobody was hurt," Neill said, as they all stood around looking at the engine lying on the floor. They resembled homicide detectives appraising a corpse—all that was missing was the chalked outline around the body. What was most surprising about the way engine looked, though, was how *unscathed* it looked; as they began to examine it more closely, they became puzzled all over again: how on *earth* could it fall and not be ruined, smashed, shattered, broken?

"Look at this," said Shade, pointing at the two feed-through housings on the face of the transition piece. The housings, which might look to a layperson like glorified hexagonal nuts an inch in diameter, were in fact expensive sleeves through which the alternator wires emerged from within the engine. They protruded about two inches from the transition piece itself, and in what can indeed only be described as an act of God, when the engine had fallen, it landed squarely on the two feed throughs, crushing them like tiny beer cans, but, in the process, causing them to act like shock absorbers. If the engine had fallen at a slightly different angle, or with a slightly different orientation, the weight of the engine would have been focused on the piston cylinder, the piston, and by extension, the magnet paddle. Neill leaned down and pushed, experimentally, on the piston face, and it moved, just the way it was supposed to.

"And look here," said Abhijit, pointing to a cosmetically ugly but purely noncritical gash in the aluminum frame that held the displacer gas spring. When the back end of the engine had followed the front end to the floor, the gas spring frame had landed on the edge of the wheeled cart, absorbing much of the impact—again, a stout part had taken the weight, entirely through chance.

"Let's . . . ," Neill began to say before being interrupted by Shade.

"Haul it up on the chains and motor it right now, and see if it works," Shade says. "The piston seems okay; let's see if it's still free. Put new feed throughs on and motor it just a little bit and see what happens."

In a minute or so of consultation, they all agree to do this; they want it up off the floor as soon as possible, anyway. Shade rigs the chain hoists, and they lift it up. Hans brings in a new set of feed throughs, and they replace the crushed ones, marveling again at their good fortune. When the power leads are hooked up and current is sent through the alternator, not only does the engine seem fine—no horrendous noises—it actually seems to motor with a bit *less* friction than it had previously. The jolt apparently "aligned" the self-aligning bearings with a force they would never have considered applying—the fall was the equivalent of whacking the engine with a dozen sledgehammers. But the current needed to oscillate the piston seems reduced. "That's what was wrong with it," says Shade. "We haven't *dropped* it enough. Let's crank it up to the *ceiling* and . . . ."

For the next half hour—as people troop into the test cell to see the results of the "disaster"—they debate whether to strip the engine down completely and inspect every part or to reassemble it and run it, in order to see if it works all right. Stripping down the engine carries the day; they were about to do it anyway, and why tempt fate? Or, at least, why tempt fate again? "Let's figure out another way of moving the engine around," says Beej. "Oh? Why?" asks Neill. He goes off to report to Bwana William that, as nearly as they can tell, everything is fine. When they next run the engine, they find no apparent negative effects from the engine drop, but the slight improvement in the bearing friction also turned out to be illusory; they had power in the 3,500-watt range, but not much more. As a result of their methodical and iterative process of elimination, it gradually became apparent where the power—the power not lost to the bearings or poor centering—must be. They decide they must open up what is certain to be a can of worms.

The piston gas spring—on paper—always looks to be a beautiful solution to a common problem in Stirling engines; it offers a direct method of establishing an operating frequency for a machine, for one thing, and an almost as direct method of mating power-to-piston to power-to-alternator: the gas spring absorbs power by making the piston work "harder," so an overpowered engine will not necessarily overpower the alternator. In this regard, the piston gas spring resembles the mechanical spring on a storm door closing mechanism: the spring absorbs the force a door takes when it is pushed open too fast for the closer to keep up.

But it is axiomatic that anything absorbing power is, in essence, using power—wasting it, if more power is what is desired. Piston gas springs are considered almost dogma in the Stirling-engine world, even though they are known in advance to use a certain amount of power that otherwise might be used to do work. The 5K team members begin to suspect that the "certain amount" of power consumed by the piston gas spring is rather larger than they had anticipated, and, with Cummins—and the rest of Sunpower, for that matter—breathing down their necks to find some more watts, they undertake to remove it.

On the one hand, this might seem to be an obvious decision—take out something that is wasting power. On the other hand, "straightforward" changes such as this have been known to bollix up the delicate dynamics of an engine so thoroughly that it never runs satisfactorily again, or runs satisfactorily only after all of the ramifications of the change are tracked down and modified themselves: there is nothing in a Stirling machine that can be changed without having effects in a dozen other places. And one of the known results of eliminating the piston gas spring is problematic: it is all but impossible that the engine will run at 60 Hz without it—55 Hz, maybe, with some tweaking and goosing, but not 60. And 60 is the frequency at which electricity is consumed. Fifty-five Hz electricity will make an hour on an electric clock take sixty-five minutes to elapse, for example, but there is a larger cost thermodynamically that Sunpower must consider.

Other things being equal, one machine operating twice as fast as another will do, by definition, twice as much work: a car going sixty miles per hour will, in one hour, have gone twice as far as one going thirty. So, if in the best circumstances the frequency of the 5K drops to 55 Hz from 60, the engine will produce 8 percent less power than it did before, and the question becomes "Is the gas spring absorbing more than 8 percent of the power of the engine?" If it is only 8 percent, there will be no net gain. If the frequency drops to say, 50 Hz, and thus costs 16 percent of the power, and the gas spring turns out to be costing only 15 percent, there is in fact a net loss, which must be offset in some way. The team members know, for example, that lightening the piston will do it, but this would mean reconceiving the piston rod scheme as well. That would mean modifying the transition piece, the displacer rod, the bearings, and a half dozen other things. If the piston gas spring is wasting one and a half or two kilowatts, they won't get anywhere unless they take it out.

But taking it out takes them back into the only place more unnerving than uncharted territory, and this is half-charted territory, where everything they do has a ripple effect. If there were a chart on which such numbers are recorded—an "All Things Equal" conversion chart— such decisions would be easy. There is no such chart. The team members must take a risk. They need more power, and they need more money, the money from the incentive payments. A working strategy emerges to try and do two things at once: make the incentives at the same time they modify the engine. From one point of view, these are mutually exclusive goals.

As an overlay to a very common process of modification, an additional problem has begun to develop, not with the hardware, but with the money. In a series of conversations that occurred as Shade and Hans and Neill began to modify the engine, Sunpower and Cummins sought to reach yet another understanding of how they should proceed, given that while they had a win with the 5K, they hadn't had a big win, and Sunpower needed more money to have a big win.

To Cummins, it seemed that Sunpower wanted to change the contract to reflect what the company had made, rather than what it said it was going to do; to Sunpower, it seemed that Cummins was taking a page from the *Sponsor's Handbook* and asking to move the goal posts: no more milestones at less than full power or outside the design specifi-cations, Cummins argued, because time was short and the program—

Sunpower—was late. The shorter the time, the more Cummins wanted a machine that was dead nuts on specification, and this included the frequency: full power at 55 Hz, in Cummins's view, wasn't a machine that would earn incentive payments. Surely Sunpower wasn't going to be fooling around trying for the incentive payments when there was so much work to be done?

What was happening was this. The *original* incentive payments had been worked out with target dates that Sunpower itself had suggested. At the time the contract was negotiated, it all looked lovely, with the incentive milestones being hit, like ducks in a shooting gallery, every month or so after the first engine run. But, with the Box alternator time and money spent but nonproductive, what Sunpower had in essence done was design two machines and build one, and very late to boot. The milestones over which Sunpower and Cummins were disagreeing were the sort of thing that earn lawyers fantastic automobiles: could Sunpower pick and choose certain specifications from within the incentive milestones as though it were ordering from a menu in a Chinese restaurant, or did it have to hit them in the manner prescribed in the contract—a prix fixe menu, as it were?

If stuck with the table d'hôte, Sunpower would likely be out of luck, and out of business: although it had had a successful first run, there were some obvious problems that needed to be worked out and some problematic losses that Sunpower needed time and money to sort out. In a theoretical sense, doing *anything* on the project that wasn't directly aimed at a production prototype—an ultimate machine—was counterproductive, because while there were things that Sunpower thought it could do to hit a milestone and pick up a check, most of those things were artificial and would retard development of the production prototype. Modifying the machine by taking out the piston gas spring and running at, say, 50 Hz but full power should, in Sunpower's view, earn it a check. Cummins's view was that this would delay the program. Cummins countered by pointing out that Sunpower *always* wanted to change the hardware instead of delivering what it said it would. Suddenly, the incentive payment setup was a lousy idea all around, and everyone started bitching. Who thought of it anyway? Sunpower wanted to forget about the incentive payments and just get the money; Cummins wanted to forget about the incentive payments and not pay the money. As the Monday meeting minutes put it, "a series of phone calls ensued."

These were battles that Sunpower had fought many times, with many sponsors. Every year that Sunpower had been in existence, William had grown more and more weary, and less and less patient, with what he saw as "management by mail" or "micromanagement," by sponsors who weren't on the scene, didn't understand the technology, and were nerveless to boot when it came to pushing the envelope. ("Micromanagement" is what all sponsors try to do, by William's definition.) Sunpower knows what it's doing, he would argue, and ought to be left alone to do it. That sponsors could be depended upon to draw the purse strings tight in order to get their way was also, in William's book, right out of the *Sponsor's Handbook*. He hated this almost as much as he hated micromanagement. But it was clear that Sunpower had to make some changes in the engine, and it could either proceed in the most efficient manner possible or futz around trying to pick up the bonus money.

Both arguments were compelling; for example, the hundred-hour milestone was tempting to Sunpower, because making the 5K prototype into a hundred-hour machine was technically doable. Sunpower could go literally for broke, fire up the machine, and try to baby it for eight days (the hours had to be consecutive, but not continuous: running the machine for twelve hours a day for eight days without taking it apart would count as ninety-six hours). If the engine didn't blow up, or blew up in the hundred and first hour, Sunpower would collect an incentive payment, but have no hardware to show for it. Even if the engine performed flawlessly for eight days, all other work on it would obviously have to stop: instead of debugging the machine, the company would be paying people to stand around and watch it. And when the hundred hours were up, Sunpower would *still* make the changes that it was obvious were necessary before the hundred-hour run, and would have lost the time used to set the machine up to run for a hundred hours as well. The production prototype was intended to be a thousand-hour-plus machine. (Much of the engine was designed with *forty thousand hours* in mind: the ultimate machine.) As William began to put it, Cummins was more interested in "freezing" the design—holding Sunpower to the hardware as it existed—than in actually moving the program along.

Cummins's point of view was equally compelling. It had paid Sunpower a million dollars for research and development: the incentives were a reward for having spent that money wisely, not a backdoor way of eking out more R & D money to do something Sunpower should have done in the first place. Abhijit and Cummins's John Bean talked. William and Cummins's Rocky Kubo talked. Abhijit and William talked; Abhijit and Neill talked to each other. John Crawford talked to everyone. The subject was—what else?—money, money, and the 5K.

The first run of the 5K, which in the tradition of Sunpower should have been an occasion for celebration, was beginning to be talked about like a bust, because it was now embroiled in niggling over the next hundred thousand dollars or so that Sunpower needed and Cummins seemed loath to spend; when indeed was the 5K team going to begin pulling its weight? That the engine had produced more electricity than any Sunpower machine had ever done, and that it had done it on the first run, was lost in carping about money and about time.

Suddenly, everything that the 5K group did was being second-guessed by Cummins, by William, by other engineers. All sorts of quick-and-dirty ideas were proposed, all manner of suggestions for "dealing" with Cummins brought up, as though Neill and Abhijit simply hadn't thought of something. "Why don't you tell them . . ." and "Why don't you try . . ." became the preferred openers in talks with Neill and Beej; Beej would generally try to explain, patiently, that they *had* said this, or *couldn't* try that, and spend twenty minutes or half an hour going over the reasons why. In the technical meeting each week, it seemed suddenly as though no engineer at Sunpower had anything to do but make helpful suggestions to the 5K team, unless it was to spread distorted information among themselves. "So your sponsor is pulling out, hm?" an engineer asked Neill in the test cell one afternoon, causing the blood to drain from his face. It was a question based on misinformation that nonetheless spread immediately through Sunpower like a bad smell, prompting Beej to generate a memo saying that reports of the death of the 5K were greatly exaggerated and that such talk was counterproductive. Neill's memo on the same subject—never written, but conceived a half dozen times a week—would have read something like:

TO:	Everyone
FROM:	The Morons on the Cummins Program
SUBJECT:	The 5K

Piss off.

For a few weeks, Neill used this standard response (and its variations) to all rumors, nonsensical nonissues, whimsical suggestions; then, instead of getting madder (Eric Bakeman told him that while Neill was smiling during the tech meetings, he still looked like he was about to explode), he thought he would get even, but the irony of the sign he posted over the 5K assembly cell was lost on everyone. "All Opinions Welcome," it read, but most people didn't notice it, so busy were they offering opinions. Since the team's most urgent priority was the hardware, it sometimes felt as though it were fighting a dragon while getting "help" that consisted mostly of grabbing onto the sword. The hardware was dragon enough.

The compromise was delicate and, in the short term, potentially suicidal for Sunpower. It was a mark of the financial situation the company found itself in that it agreed to it in the first place. Incentive payments would be made for a series of runs at full power, period. Another set of incentives would be put into place for runs at *six kilowatts* power and full spec—60 Hz and 25 percent efficiency. The rationale here was breathtaking: if Sunpower could make five kilowatts at 52 Hz, then the same machine could make six kilowatts at 60 Hz, and Cummins would be getting a bonus of sorts itself: a more powerful machine without a full-blown redesign.

Sunpower could earn itself a certain amount of money, in other words, by getting up to five kilowatts, but it would then have to *help* Cummins move the goal posts and make the 5K turn into a 6K— without spending all of the incentive money. Otherwise, its people would be back where they started, at best, and out of business, at worst. The really nervy part of all this was the need for a parallel series of modifications that would take the machine up to six kilowatts for forty- and one-hundred-hour runs. If they ran into something unforeseen, they would be in the position of having spent the incentive money on an irrevocable path that would leave them no way to get more incentive money. Pessimists call this eating the seed corn. William saw it as— what else?—an opportunity. Since the only thing that sponsors ever seemed interested in was more power, it was the only thing, really, that Sunpower had to bargain with. The morning after all this was arranged, everyone got to work bright and early.

The team members started with the piston gas spring. Some ideas are hugely complex to implement, but easy to conceive. Removing the piston gas spring looked to be the opposite: it had taken many weeks of engineering to design it; it took Dave Shade and the Bridgeport milling machine a single morning to make it disappear. By cutting holes in the half-inch-thick steel plate that formed the "bounce space" for the piston, the piston gas spring vanished, an annulus turning into metal chips. If things went right, Sunpower would earn an incentive by run- ning the machine at a lower frequency, but five kilowatts, and would have some breathing room. Hans and Shade assembled the engine, and a by now blasé team assembled to watch it run; in two dozen engine runs with this machine, all had gotten used to flipping a switch and starting it right up. They even made plans for Hans and Shade to stay on into the evening if they got up to full power right away, so that they could run the engine for eight hours and pick up a check.

The engine didn't start.

For three and a half hours, they tried everything they could think of, from using sophisticated electronic testing equipment to jiggling wires, and the engine simply wouldn't start. Neill asked Shade whether he had put in all the necessary parts; Shade stared at the test cell ceiling, and everyone else stared at Shade as he assembled the engine in his head. Hans went to the assembly cell to see if there were any obvious parts left over. Larry Haas, the chief technician (and a licensed electrician) came in and painstakingly went over the electrical grid connections, checking fuses and phases. Scott McDonald, an electrical engineer, came in and painstakingly went over hundreds of resistors and logic boards. They spent a full hour determining that the kill valve was closed (Shade pressing his ear to the pressure vessel and listening for it to click), which it appeared to be. And again and again, when Abhijit hit the starter, the engine refused to cooperate. When they left that evening, it was not to go home for a restful night.

The next morning, they took the engine apart, looked everything over, assembled it again, step by step by step. No soap. A last-resort strategy—Neill staring at the pressure vessel and calling the engine a piece of shit—was also unsuccessful. (Shade and Hans had both tried this the previous day.) If thinking hard produced energy, rather than consuming it, the half dozen souls in the test cell could have lit the Eastern Seaboard. And nothing they tried worked. The frustration level was such that Neill started throwing people out of the test cell when they came in to offer advice; all opinions were no longer welcome.

They finally figured it out through a method that, while not taught in engineering school, reflected exactly the sort of resourcefulness that made Sunpower such a compelling place to be around. As Beej flicked the starter over and over, Shade and Neill stared at the momentary flickers of data that appeared on the computer screen, looking to see if in the transient numbers there was something obviously different from all the previous runs.

There was, although it was bizarre enough that they could be forgiven for not having considered it: the phase angle of the pressure in the back end of the engine was suddenly showing up as a nonsensical number, meaning that the helium, instead of ebbing and flowing in phase, was colliding with itself in an unexpected turbulence. "Uh oh," said Shade, who suddenly remembered having seen this in another Sunpower machine, years and years ago.

The engine had been designed with the piston gas spring, which meant that the backward movement of the piston, or, more important, the pressure wave in the helium that it produced, was captured by the steel plate on the alternator. When Shade milled out the steel plate, he eliminated the gas spring, but not, of course, the pressure wave. And in a curious confluence of circumstances, the round steel puck on the *displacer* gas spring (which shouldn't have had anything to do with the *piston* gas spring) happened to be exactly the right distance away from the piston that a "feedback"-springing effect occurred, the pressure wave from the piston colliding with the pressure wave from the displacer spring the way a wave of water rolling back from a beach clashes with the next incoming wave and makes a splash. An engineer who set out to try and cause such an event would have to spend weeks and weeks doing so: it is an effect that is subtle in the extreme, a sort of master's thesis from hell: where, to the nearest millimeter, does feedback springing occur? It was the equivalent of dropping a coin from the top of a building and having it land, edge up, in a crack in the sidewalk.

William, on being told of this, remembered Sunpower's previous experience with the phenomenon as well, and offered the solution— change the displacer gas spring slightly, and thus change its pressure wave without changing its springing capacity. Neill delayed telling William what they had discovered long enough to get Shade to do exactly this, so that he was at least able to say that the parts were in the shop. William, who hadn't been happy with the compromise in the first place, was so anxious to have the team start work on the "real" modifications that he cited this event as proof that the whole enterprise of aiming for incentive payments rather than engineering was a waste of time, and he told everyone so, getting into an argument with Rocky Kubo at Cummins when Rocky alluded to Sunpower's throwing good money after bad.

William, in fact, was on the phone shouting when the modifications to the displacer spring were finished and the engine started right up and ran at 50 Hz and five kilowatts; he came into the test cell and greeted the news that the machine was producing five kilowatts—the design specification, for the first time ever—with a curt "Good," which infuriated Neill, who hadn't known about the phone call. (There were days when a phone call from Cummins could make even Abhijit grouchy.)

Two days later, when they ran the machine for eight hours at five kilowatts, and earned an incentive payment crucial for the company's financial survival, William was already pressing Neill for the work that would transform the machine from a 5K to a 6K: no one wanted a five-kilowatt machine any more, so not very many people were thrilled that they had one, even though they had never had one before. Even though *no one* had ever had one before.

The morning after the eight-hour run, everyone was still so cranky and out of sorts that no one was in a mood, really, to appreciate just what had been achieved.

Chapter 12

Men in Suits

The Big Boss and three other White Males in Suits visited
and saw everything we have to offer. The Boss is a smart man
even if he is interested in Stirling engines, and Crawford says
that when he sees a way to make a buck, he'll be back.

—*from the minutes, July 1, 1990*

I had a nice conversation yesterday with the Director of
Global Change. . . .

—*Lyn Bowman at a Monday morning meeting*

TO MAKE THE LEAP from five kilowatts to six, the 5K
group had to come up with something that would,
first, lighten the piston considerably and, second, use
less power to overcome the friction in its bearings. It also had to do this
with very little money and very little time—not to mention while its
members were in a spectrum of moods that ranged from depressed to
out-and-out foul.

Late in the debugging process, when they thought that the 5K
had to be only a 5K, they had conceived a new bearings scheme that
would take advantage of an innovation made by the Stirling-cooler
engineers at Sunpower, an innovation that looked to be a very straightforward retrofit onto the existing 5K design. It was, in fact, this scheme
that Sunpower considered one of its aces in the hole when it was
jousting with Cummins over the incentive payments.

However, when the goal posts were moved to six kilowatts, the
strategy became at once more focused—they would *have* to use the
cooler group's invention—and more problematic: they weren't sure
how. Robi Unger, the cooler group's bearings maven, was "borrowed"

from the Cucumber cooler project to put the 5K on the fast track as far as its bearings were concerned.

That Robi was available to be borrowed (meaning he would bill some of his time to the 5K project, rather than to the Cucumber) at all was a hint as to Sunpower's financial position. For the engineer largely responsible for a hugely important technical advance not to have any-thing (funded) to do might seem odd, but it was true, the result of a classic Sunpower financial bind that had only a peripheral connection with the technology. Robi was available because the most important cooler sponsor—code-named Cucumber, in deference to the sponsor-ing corporation's passionate desire for secrecy—had decided, after Sun-power demonstrated a fantastic technical success, that it wasn't really interested in the device Sunpower had made for it after all, and not much more funding would be forthcoming. This was bad enough; combined with what was becoming apparent about the Cummins sponsorship of the 5K, Sunpower found itself facing the prospect of losing two million dollars or so of funding that it had been counting on. Worse still, both the Cucumber and the Cummins monies were stepping-stones—or had been counted on to *be* stepping-stones—to larger, longer-term funds. Worst of all, Sunpower had spent a notable portion of Cucumber and Cummins money that it didn't have, and now looked as though it might not get after all.

The 5K team was in the position of finding watts and losing fund-ing; the Cucumber team was in the position of finding lower and lower temperatures and losing funding. Technical success, financial catastrophe: *why?*

As far as the solar 5K project was concerned, Sunpower was find-ing itself in exactly the position that William warned about when he said that hitching the company's financial well-being to a corporate behemoth left Sunpower with a lack of control over its destiny. (Wil-liam is not above saying, "I told you so.")

Cummins's long-term plans for the 5K were fairly certain: it was going to pursue the research. In the short term, however, Cummins was susceptible to the lure that every company in a capitalist society is susceptible to: playing the capitalist game with someone else's money, preferably the government's. The U.S. Department of Energy had pub-lished a "request for proposals" for a solar thermal program that it would fund on a cost-sharing basis with a company willing to develop the technology. Cummins, which had been playing largely with its own money before, saw an opportunity to extend the development of a project it had already invested in, but using government funds, and it submitted a proposal to DOE, with itself as contractor and Sunpower, Thermacore, and Cummins's own subsidiaries LaJet and McCord as subcontractors. Although Cummins maintained that it was committed to solar Stirling technology with or without government money, the government money was too good to pass up: the economy was in recession, and diesel engine businesses are extremely vulnerable to economic downturns—the engines are so durable that rig owners patch up, rather than buy new, when money gets tight.

Cummins would weather the downturn, but needed to watch its money—in corporate America, this means "cut out R & D," especially if government money is available.

In a world that moved in tune with the DOE, this would make little or no difference to anyone. But DOE moves with such impressive deliberation that time seems to stand still: it thinks nothing of simply delaying a proposal review for, oh, a *year*. An absolutely minimum level of self-protection for a company's financial health would declare that government money—especially DOE money—could be only icing, not cake. (William, and Sunpower's board of directors, had said this many times.) Wait for DOE, and go broke doing it. Sunpower's heat pump program, on which Jarlath McEntee and Gong Chen had spent two years, was the epitome of DOE programs: when technical difficulties developed, the DOE program managers decided that Sunpower should do an extensive series of tests and computer simulations and then write a report. Every time Jarlath and Gong thought they were finished with the report, the DOE would request another report. To spend the productive years of one's life writing reports for a government agency is few people's idea of engineering, certainly not Jarlath's and Gong's. But once one leg is in a government trap, it can be expensive to chew it off: without the money that DOE was paying Sunpower for Jarlath and Gong to write reports, Jarlath and Gong would be out of work. And abandoning a project in a snit can mean being dealt out of the next hand of government money—Sunpower will never work with NASA again, both because Sunpower doesn't want to and because NASA won't want to, either, Sunpower having burned its bridges on the 25K debacle.

And now the solar thermal program—what had not much earlier been seen as Sunpower's ticket to the private sector—was turning into a government project, on a glacial government time scale. (When the end of the world comes, the DOE will be the place to be: it will take an extra year.) When six months had passed after the DOE application deadline, Sunpower took note; when nine months had passed, Sunpower got nervous. When Cummins mentioned to Sunpower that there might be a fallow period between the 5K project as currently constituted and the beginning of DOE funding, Sunpower saw only one thing to do: panic. When the DOE sniffles (and thus decides to lie down for six months), Cummins might catch a cold, but Sunpower develops lobar pneumonia and rickets besides. Without Cummins's money, there would be a lot of people at Sunpower with work to do, but no one to bill it to. The DOE funding—the shorthand name for the government program was "Mission III"—that Cummins had decided it would depend upon would leave the 5K in the worst possible position: work to do, no money to do it. Mission III, as it was presented to Sunpower, was "Mission: Impossible."

And then the Cucumber program vanished. When Cucumber Inc., called John Crawford and told him that it wasn't quite ready to fund Sunpower in the manner to which Sunpower had become accustomed, the whole principle of sponsor-funded research took a beating on Byard Street. So did a large number of engineers, technicians, machinists, and office personnel. What made Cucumber position seem so . . . *unfair* was simple: it is one of the most fabulously wealthy and successful corporations in America, a company that makes more in one day's revenues than a half dozen Cucumber programs cost in a year. The cost of the Cucumber program, in fact, amounted to less than one day's *profits* for Cucumber, Inc. And the initial Cucumber program was an unqualified success.

But wealthy companies get wealthy in large part by not spending money on anything they don't have to, and the Cucumber was something that Cucumber, Inc., decided it didn't need any more of at the moment, although it still held the license. No one else had one, so it provided no competitive advantage. If someone else came up with something similar, well, Cucumber, Inc., would be way ahead of the game. In the meantime, Cucumber, Inc., put the Cucumber on the back burner or, rather, on ice.

Now it was John Crawford's turn to get helpful suggestions from everyone as to what to say to the sponsor to make it cough up the money. The loss or suspension of money from these two sponsors led to one of the worst financial episodes in the company's history, one made more bitter by the presence of so much working hardware. Make something of value, and it will make money, or so the original Sunpower argument went. Its employees can be forgiven for feeling they had done their job, or most of it. But the world wasn't noticing.

This is what drove Sunpower to the Sunpower Massacre, a pruning of the company's payroll with a chain saw. There is no good way to lay off people who have given excellent value to a company, but the way Sunpower went about it—*had* to go about it—was arguably more traumatic than most. Put simply, those who were of most *billable* value stayed: the more hours an employee could count on billing to a funded project, or to one where funding was in the pipeline, the more likely that employee was to keep a job. Wrenchingly, the company laid off several engineers whose only sin was working on projects that were underfunded or not funded at all: one of the small-engine designers lost his job,and the other was cut to half-time (and put to work on other projects); Barbara Fleck, who had been "borrowed" years ago from the cooler group by the 5K team and who was on maternity leave to boot, was out of a job. (Shortly thereafter, another engineer on maternity leave, Mary Fabien, was also laid off, when the project she was working on ran out of money. Although the rationale was perfectly businesslike, having a baby in the Sunpower family started to look like a bad career move.) Joerg Seume lost his job, as did Eric Bakeman and Morgan Mitchell—the computer mavens—and Mark Labinov, whose specialty, an encyclopedic grasp of higher mathematics that he used to update the SAUCE computer analysis program, was suddenly expendable. The ranks of machinists and technicians were thinned out, and those who kept their jobs kept them at half-time, with a pay cut besides. The unremittingly cheerful and unremittingly efficient accounting and business staff was cut by two-thirds; even the chocolates passed out when time cards were turned in were eliminated. Everyone who was left took a temporary pay cut in the Sunpower tradition: the more you made, the bigger the cut. And even this bare-bones operation was vulnerable, because the company couldn't survive without getting Cummins/DOE money,and getting it quickly. With private sponsors, Sunpower had often billed in arrears—it would start work on a project while the paperwork was being attended to, and could count on being reimbursed when the money began to flow. This was not possible, however, with government money: a contract had to be signed *and* approved by DOE before billing could start, and there could be no "back billing," something the government refers to as "criminal fraud."

With Cucumber, John Crawford began intensive renegotiations with the people at Cucumber, Inc., trying to change their minds, while at the same time trying to "shop" the Cucumber to anyone who had ever made anything remotely cold. He and David Berchowitz took a Cucumber-cooled mini-refrigerator to an international conference in

Montreal and served soft drinks from it to engineers from around the world. Crawford and William and Lyn Bowman began working the phones, beating the bushes, stirring the waters; they sent out flashy color brochures with David Berchowitz's smiling face next to a Cucumber covered with ice; they sent out press releases and photographs of a Cucumber next to a can of orange soda, to indicate its compactness.

And in a move that enraged Neill Lane, they dismantled Todd Cale's drafting computer for an afternoon so that they could use it as a prop for the Cucumber, and in doing so, placed a Cucumber, with its ring of magnets strong enough to stop a clock, on top of the magnetic computer hard disk containing every single design drawing in the 5K's long and tangled history; it made for a great picture to send out to computer companies, but one that Neill Lane couldn't appreciate. "Why didn't they use their *own* computer," he fumed. "They don't have any drawings for that little piece of shit, anyway." No drawings were lost, but to Neill it was a sign that the 5K was being taken for granted yet again. But the Cucumberians couldn't be troubled: they had press releases to get into the mail.

This sudden flurry of publicity annoyed Cucumber, Inc., which considered it a violation of the contract, which specified utmost secrecy: the company felt it had contracted for a machine, while Sunpower interpreted the contract as one for a "specification"—to provide so many watts of cooling with so many watts of power consumption; anything outside Cucumber, Inc.'s specification was, in Sunpower's opinion, its to try and sell. Cucumber, Inc., though, hadn't become Cucumber, Inc., by sharing technical information with a curious world, but rather by being tight-lipped and litigious. One afternoon found John Crawford at the front desk, reading a letter from Cucumber, Inc.'s legal counsel that threatened all manner of horrible things if Sunpower didn't *shut up* about the Cucumber and quit trying to drum up interest in it. Crawford set it aside to handle later. The next letter he opened began, "I read with interest about your small Stirling cooler in the current issue of *Popular Science*." Beat the bushes? Sunpower had no choice but to pull out all the stops. Sunpower did "cease and desist" clouting the Cucumber around, but only grudgingly; Cucumber, Inc., insisted it was still very interested in the technology, but not right at the moment, and Sunpower wasn't doing itself any favors by antagonizing a once and perhaps future client.

It was in this atmosphere that the 5K was supposed to metamorphose into a 6K. Not that the team members felt any pressure. Not at all. They simply had to have some heroics. The old piston rod was, de facto, simply out, even though they were well on the way to a new bearings scheme. It was simply too heavy. So they pushed the envelope in Sunpower fashion, and moved full speed ahead on two huge changes at once, the sort of changes that a real company would spend a year doing experiments for. Sunpower had, at best, a few weeks. And while the people in the 5K group (they resolutely kept their name, despite needing another kilowatt) felt sometimes as though they didn't get any respect, they were perfectly happy to use technical innovations from Cucumber in the modification of the engine. The innovations were just too clever to ignore.

Stirling coolers have been around almost as long as Stirling engines have—since the nineteenth century, at least. Stirling coolers based on a crank design had, of course, the same limitations as engines with cranks, principally sealing problems. But free-piston Stirling coolers, with a sealed pressure vessel, are lovely machines, simple, inexpensive, fully the equal of their chlorofluorocarbon-pumping counterparts. A Stirling cooler circa 1975, for example, was a solution in search of a problem. Then came the information that CFCs were damaging the earth's protective ozone layer, and one would have thought that Sunpower was sitting on a pot of gold: a non-CFC-cooling device, within a percentage point or two of the efficiency of vapor compression devices, manufacturable for nearly the same cost even as a prototype (without taking into account economies of scale). No wonder General Electric—the largest manufacturer of refrigerators in the world—was interested enough to fund some Sunpower R & D. There was some wonder when GE lost interest.

The "GE cooler," as it is known around Sunpower, is, at first and second glances, an absolutely perfect machine. It is about the size of a conventional refrigerator compressor; it uses the same amount of electricity; it provides as much cold; its only expensive parts are the iron-neodymium magnets in its alternator. And it uses environmentally benign helium, rather than ozone-destroying CFCs.

On the third glance, questions can be raised having to do with manufacturability and durability, both of which are unproven for a Stirling machine. Conventional refrigerator compressors are routinely guaranteed for a decade and, in practice, last three times that; they are manufactured on assembly lines all over the world, because they are so standardized as to be a commodity item.

Conventional compressors last so long because they are filled with oil: a piston about the size of a baby-food jar rides on a small crank in a container of lubricant, and metal parts bathed in oil will run for a long time. A Stirling cooler, by contrast, uses no oil—can have no oil, because oil would foul its heat exchangers. So what sort of running surface, metal on metal without oil, can last for thirty years?

The first attempts at Sunpower to answer this question date from the late 1970s, and provided some of the basis for Robi Unger's later idea. Spinning the piston, William decided, would produce a "spin bearing," or a thin film of gas between piston and cylinder as a lubricant—the piston would "float" inside a jacket of helium produced by the spinning motion of a piston. This is, in fact, the way oil works as a lubricant: in a typical metal-to-metal connection called a "crank journal," the motion of the crank actually produces a wave of oil in front of it, so the metal parts simply don't touch. Sunpower actually developed some machines with these spin bearings, but never pursued them to a point where they were workable and reliable in practice.

Then came the Cucumber project, which had technical specifications that seemed daunting at first. One of the more daunting was a "mean time between failures" (what a layperson might call "life expectancy") of 100,000 hours, or eleven and a half years of continuous operation. There exist no materials of any kind that can rub against each other sixty times a second for eleven years without some sort of lubrication. Robi Unger, with David Berchowitz, turned the "spin bearing" into a "gas bearing"; instead of relying on the spinning of the piston to create a film of gas between a moving part and its cylinder, the Cucumber gas bearing used cleverly placed "gas pads" around the piston and displacer, fed by a check valve and tube arrangement, to hold the piston forcibly away from its cylinder, something like the way a hovercraft hovers. The movement of the piston itself provided the pressure used to force helium through the gas pads and keep the piston from touching its cylinder. No touch, no wear. No wear, no wearing out.

A clear benefit of gas bearings is the virtual elimination of friction and, hence, an improvement in machine performance, since no power is consumed overcoming friction. With gas bearings, a Stirling-cycle refrigerator would seem to run forever, or at least a very long time. The Cucumbers were the first machines to use gas bearings, and although Sunpower hasn't run them for eleven and a half years, it has run them for thousands of hours, and they work.

Robi Unger and Nick van der Walt spent several weeks retrofitting one of the GE coolers with gas bearings, and the results were the same: an efficient machine with no wearing parts. A Cucumber, although it was designed for a computer-specific application (one of the limits to the speed at which computers work is heat: cooling a computer chip to low temperatures permits faster computer speeds), provides enough cooling power for a small refrigerator; the GE cooler-sized machine is capable of cooling the largest refrigerators manufactured today—for that matter, it is stout enough to cool supermarket coolers and freezers, a huge refrigeration market in and of itself. One can almost see the path beaten by the world to Sunpower's door.

So, two machines sit in Sunpower's lab these days that represent a reconception of the common refrigerator so unique that the very uniqueness has, alas, become a liability; a third machine, nearby, is a marvel technologically, but more of a marvel as an example of how big business in the United States works: the two machines represent a leap forward to what one would think would be the Holy Grail of refrigeration technology—CFC-free cooling devices. It is the third machine's story that is so poignant.

The third machine marks a step backward, a Darwinian survival of the less fit: it is a CFC-free Stirling machine retrofitted to *use CFCs*. If you can explain this, please call William Beale. This state of affairs has flummoxed even him, a man who thought he had heard everything about how difficult it is to bring new technology from the lab to the consumer. It is not the first time that it seems as though the inmates have taken over the asylum.

William did get a glimmer of how men in suits view CFCs from an ominous precedent of several years back, when the U.S. Department of Energy (at Sunpower, they would say, "Who else?") solicited proposals for computer simulations of non-CFC refrigeration devices. At the time, Sunpower had not only a computer simulation but the device itself—the sort of thing that one might think the Department of Energy would be interested in knowing. In fact, DOE was not: just the simulation for now, please; hold the hardware until you've tested the theory of whether or not such a device is possible. Sunpower tried, to no avail, to entice DOE into *looking* at their machine, but DOE covered its eyes. ("You can bring it here if you *want,* but I'm *not* going to look at it," is what a DOE program manager actually said.)

So Sunpower applied for the simulation contract and was awarded a DOE research grant to do a computer simulation, in essence earning itself money to demonstrate, via computer software, the possibility of the existence of a machine it had already made. Mary Fabien, the engineer who performed the study, did much of the work, in fact, within *sight* of the very machine she was ostensibly determining the potential existence of. It was like driving to Washington to deliver a computer simulation of the potential existence of a wheeled personal transportation device powered by an internal-combustion engine, or using a laptop computer on a coast-to-coast flight to study the possibility of an airplane.

To call the DOE bizarre is not entirely fair: to say that it believes in proceeding methodically and systematically is closer to the mark. The results of the study, which Mary Fabien and David Berchowitz presented at a two-day conference in Washington, D.C., were unequivocal: yes, indeed, the computer simulations predict that a Stirling-cycle cooler can be made that is competitive with conventional vapor-compression technologies. (At various times during this project, Sunpower decided against bragging that it already *had* such a machine, for fear that this might make the data seem tainted or give DOE an excuse to do something like make them start over; Mary and David didn't mention it once, in Washington, although David did allow that he thought Sunpower could come up with something pretty quickly. They in fact had a Cucumber in the *car*.)

Having proven the possible existence of a machine that they had already made, they thought it reasonable that the next step—even for the government—would be to make one. Sunpower would love to get government money to make a new batch of Stirling coolers, if only because it has learned so much in the several years that have elapsed since it first made one. But no calls from DOE were forthcoming. Instead, Lyn Bowman happened to call the director of global change.

The director of global change, it hardly seems necessary to mention, is an employee of the U.S. government. (*"Great* job title," said everyone at Sunpower when Lyn reported on his conversation. "Gee, was the director of the *universe* out of the office?" asked Ben Wilkus.) Specifically, the DGC works in the Environmental Protection Agency and is charged with the responsibility for monitoring, well, global change—exactly that. The DGC listened with some interest to Lyn's rendition (honed in hundreds of these cold calls to places all over the world in an effort to drum up business for Sunpower; Lyn would be *fantastic* in telemarketing) of the Cucumber program and the GE cooler, and let it be known that though there was nothing on the horizon for a CFC-*free* device, the EPA was about to float a proposal for an *HCFC* machine. Was this up Sunpower's alley? "Give us some money, and we'll see what we can come up with," said Lyn. The director of global change was, unfortunately, unaware of the study just funded by his counterparts at DOE; HCFCs were what the EPA was aware of.

Hydrochlorofluorocarbons, or HCFCs, are what most CFC-based

industries believe will be the only affordable substitute for CFCs; although HCFCs are still chlorofluorocarbons, they are thought to be only about 5 percent as harmful to the ozone layer as CFCs are: if all the CFCs in the world turned magically into HCFCs overnight, the thinking goes, the problem would be 95 percent solved. (The fine print in this argument is much more complex, but this sort of statement looks marvelous in the publicity brochures.)

Vapor compression technology fans (these would be all of the present-day manufacturers of CFC-using devices) like HCFCs because they work very much the way CFCs do—they are familiar. They also tend to be made by the same companies that make CFCs, so the supply systems are already in place. That HCFCs are incompatible at present with conventional vapor compression technology is a problem that the U.S. government is seeking solutions to, in essence on behalf of Du Pont, GE, Whirlpool, and their clouty counterparts, even though HCFCs, not to put too fine a point on it, simply don't work in conventional compressors: HCFCs are chemically incompatible with the lubricants in conventional compressors—incompatible, in other words, with the very aspects of vapor compression that makes it a technology worth saving: durability (thanks to lubrication) and manufacturability.

This is not exactly how GE et al. put it; virtually every major compressor manufacturer in the world has been through Sunpower to see the GE cooler, and all of them say the same thing: the compressor business is so cutthroat that it would be financial suicide to do something different. A man in a suit from the second-largest manufacturer of refrigerator compressors in the world told John Crawford that he had *always* thought that Stirling refrigeration was the way to go, but that his margins on compressors were pennies and tenths of pennies, and there was simply no room in the budget (of a $350 million company) to do anything so uncompetitive as pursue technology that its competitors were not. The big boys are heading toward HCFCs mostly because the big boys are heading toward HCFCs. "I wish I could help," Mr. Big told Crawford.

The EPA compressor that Sunpower has developed is interesting for two reasons; it is the first—and, for now, the only—compressor compatible with HCFCs that looks to be durable and manufacturable and economical to produce. Also, it has been developed by a company that finds the technology to be unequivocally purposeless. Right next to the EPA compressor is the GE cooler, which uses no CFCs of any kind. If the problem is CFCs, which one would you pick?

This line of reasoning has gotten Sunpower almost nowhere; the

EPA compressor is not a Stirling machine at all; it is a vapor compression machine that uses a linear motor to drive a piston that compresses the working fluid. Because the Sunpower people, over two decades, worked to develop machines that use no lubricants, they were in a good technological position to develop such a compressor, although they wish they didn't have to. And the very fact of their technical success with the EPA compressor bodes ill for the CFC-free Stirling cooler— the huge refrigeration industry might be inclined, or forced, to retool once in the near term, but it will never retool twice, and the U.S. government, with the help of the director of global change, wants to abet them in retooling for HCFCs. It is not the first time that Sunpower has gotten into bed with a sponsor it would prefer not to be seen in public with, but what Moliere once said about writing is also applicable to R & D. First you do it for love, then you do it for a few friends, but finally you do it for money.

Sunpower's most rending crisis prior to the Massacre involved just this issue. There was a design competition for a long-lived, quiet, efficient generator set, the sort of portable generator that could be set up somewhere and simply refueled from time to time. The specifications were rather more stringent, but the idea was not much different from the original Sunpower idea, that of the magic box. But the sponsor of the design competition was the U.S. Department of Defense, and the debate at Sunpower was over whether or not to take DOD money.

The arguments for taking the money would be persuasive at most any other place on the planet: this was not developing a bomb or a chemical weapon, after all, but rather a portable generator. William even argued that their taking DOD money for such a benign project as this would leave correspondingly less DOD money for making bombs. It was not a military project, per se, just one where the money happened to have been budgeted to a government agency that Sunpower found to be de facto odious. The contract would, further, allow Sunpower to benefit from the R & D work itself, by learning more about making certain kinds of machines. Sunpower debated this issue, which became a sort of symbol for what the company would—and wouldn't—do for money. Sunpower took the money; some of its best and most popular engineers then left, some of them on less than good terms. The source of the money was one difficulty; the way the subject was handled was another. (William told one engineer in a fit of anger, "I don't owe you guys a damn thing," a statement that William has yet to live down.) Some of the engineers who left are still "consultants" in one form or another for Sunpower; one went on to form Stirling Technologies,

Inc., the "other" Stirling-engine company in Athens, in order to try and develop and sell the Third World engine on which Sunpower had spent so much time and effort.

But the issue of where the money comes from had been broached, debated, and, over the years, second-guessed. In a practical sense, Sunpower came out very well from the DOD project, even though the money dried up before too long. (In fact, the SHARP engines that led to the Cummins program were developed for this DOD generator set; without the DOD contract, there might well not have been a SHARP; without SHARP, there would not have been a 5K.) But the irony of using DOD money to develop technology intended to make a more peaceful world was not lost on anyone at Sunpower, and the EPA compressor project could be seen as something in the same vein: is there such a thing as "moral" technology?

If the EPA compressor had required something more dramatically immoral (by Sunpower standards) than HCFCs—uranium, say—the question would have been harder to answer. As it was, Sunpower used the same arguments it had used for the DOD money: the EPA compressor money would help Sunpower survive, develop other technologies, and draw a small amount of money out of the hands of mischief makers. But it does put Sunpower in bed with companies and technologies it would rather not sleep with.

In a heated discussion once with Abhijit Karandikar about sponsors for the small engine, Neill Lane captured this attitude succinctly. He said, "*If* I had a hundred million dollars, I might go to China, build a factory, *give* them the engines and feel good about having done my part to make their lives better; I could go right to heaven. But *if* I *don't* have a hundred million dollars, I'm *not* going to go to China and bankrupt myself in the process: I'm going to go to Mitsubishi. Whoever has the gold makes the *rules*."

And so it was with the EPA compressor and, in a subtle way, with the Cucumber as well. Without money from Cucumber, Inc., the Cucumber was again a solution in search of a problem. David Berchowitz and Lyn Bowman spent many months developing a sort of Super Cucumber, one that cooled down to superconductor temperatures, since superconductor research is attracting lots of research money these days. There are some unique and certainly moral uses for superconductors; there are also plenty of odious uses (the Department of Defense is fascinated, to the tune of many hundreds of millions of dollars, with superconductors). That one of the theoretical uses for superconductors—transporting electricity with little or no resistance and, hence, little or no loss—would to some degree be obviated by a successful solar program (you can't waste what you get for free) is, in the practical world of Sunpower, counting angels on the head of a pin. There were dark days in 1991 when a few people at Sunpower worried that some really *creepy* potential sponsor would drop by—a Third World dictator, some black-bag spy agency, a slave-labor broker—and Sunpower would lunge for the money like a trout.

Indeed, the day that the Hughes Aircraft Company came by to see the Cucumber marked to some people at Sunpower the low point, because Hughes makes William's old nemesis Curtiss-Wright look like Mother Theresa. William compared Hughes in the Monday morning meeting to "the sand worm" in the science fiction novel *Dune,* an absolutely efficient killing and eating machine, and a company that has made some genuinely scary weaponry for the DOD over the years (Sunpower half expected the Hughes people to have their briefcases handcuffed to their wrists). William cautioned everyone to keep anything remotely secret under tight wraps, and Hughes was permitted to look at the Cucumber, but not to touch it. On the bright side, that Hughes was even in the building showed how democratic Sunpower had become.

So in the pantheon of projects possible when one needs money to survive, work on the EPA compressor is probably not a mortal sin. Nick and Robi and Andy Weisgerber made it operate very well, to the point that Sunpower bought a couple of refrigerators and put EPA compressors inside them. While it wasn't trivial what they had done, it was remarkable how the EPA compressor benefited from so many years of Sunpower experience: it was the closest thing to a real retrofit into existing technology that Sunpower had ever done. Using the carcass of a conventional compressor for a pressure vessel, and indeed using as many parts from a conventional compressor as possible, the EPA compressor was up and working when EPA announced, with fanfare that made the *New York Times,* what sounded like a Sunpower technical fantasy come true: a "super efficient" refrigerator *contest,* best company wins, to design a twenty-first-century refrigerator. Sunpower's phones rang off the hook, making John Crawford cranky to the point of surliness, as well-wishers called up to tell John to look right there in the *Times: Three hundred million dollars!* Would Sunpower "enter"?

Crawford—and everyone else in the refrigeration industry, for that matter—had known about this noncontest for months. Although news coverage made it sound as though a visionary inventor in his garage could snap up $300 million and save the ozone layer in the process—

earn money *and* a spot in heaven—the "contest" was nothing of the kind, and Crawford was tired of hearing about it. What EPA called a contest (it actually said "competition") was pretty much a gift to some as yet unselected refrigerator manufacturer to design a refrigerator that was, yes, more efficient, but more efficient in ways the EPA decreed made the most sense. Anyone was permitted to enter a proposal, but part of the rules was demonstrating a capability to manufacture millions of refrigerators. Further, the money would come in the form of a rebate for each refrigerator sold—it was not R & D money at all, but a government subsidy to the manufacturer to compensate for the extra expense of making a refrigerator more efficient. Further still, "winners" would be judged on an EPA point scale, where superinsulation, for example, was worth twenty points, but a more efficient compressor was worth five. The best that Sunpower could hope for from this program would be hooking up with the winner as a subcontractor, and this hope was fairly dim: any corporation big enough to win was also big enough to have invested hundreds of millions of dollars in manufacturing capacity for conventional compressors, and no one was likely to spend the rebate on retooling.

So Robi Unger was free to be borrowed by the 5K, in the search for a spare kilowatt. It is not automatic that machines have 20 percent "extra" power hidden inside them: the gross power of the 5K was always intended to be about 6K, leaving some room for losses and, more important, room for the engine to run its own control system—the motor that moved the concentrator assembly, the motor on the pump that circulated water through its cooler, the electronic controls. It is also wise with a solar machine to have some measure of extra capacity in the alternator, so that if for an unforeseen reason—an electronic failure in the controls, for example—the alternator has to absorb more power than is nominally intended, it can do so, rather than being overpowered. Some glitch in the heat pipe control could make the hot end hotter than it should be, and in the four seconds it would take the plastic-film mirrors to defocus, the alternator might be extended beyond its capabilities. With an alternator operating at the edge of its capacity, this will make the engine run away. A margin was thus included from the first.

If the internal losses could be minimized—primarily frictional losses—the 5K would also be getting "free" power, power the designers had considered lost to the mechanics of the design. There were several hundred watts, at least, in the original bearing scheme, and several hundred watts in the sealing rings on the piston. This was where the gas bearings would come into play.

But the piston rod had served a crucial purpose, supporting as it did the magnet paddle. What could replace it? The first idea—another ringlike "spider" at the out end of the magnet paddle—looked promising for a while, but didn't pan out. Another possibility was geometrically complicated to conceive, but had the potential of solving a pair of problems at once. By moving the displacer gas spring *inside* the inner laminations—where the piston rod and its bearings once were—one could attach a wide aluminum ring to what was once the base plate of the alternator: this ring would ride inside the edge of the magnet paddle, providing support and, as important, a perfect surface for the gas bearings: the magnet paddle would "float" on this ring, held there by helium pumped through the pads. For the other end of the magnet paddle, a similar ring was sketched out, so that the most crucial part of the entire assembly would seem to rest handsomely on two wide, round aluminum rings: perfect support.

This, however, necessitated a change in the piston itself: most of the frictional losses in the piston came from the carbon-fiber sealing rings that rubbed back and forth on the piston cylinder: they sealed the work space from the back-end space all right, but did so by rubbing tightly, causing textbook friction. So the piston was reconceived as a "clearance seal," rather than a ring seal: the piston grew a long skirt and gave up its rings.

A clearance seal works by having a long, close fit between piston and cylinder—long and close, but not touching. Gas bearings on the piston would keep it from touching, in fact, so there would be literally no friction.

There would still be a loss, however, this time from leakage from work space to the back end—a clearance "seal" is no seal at all, just as narrow a crack between two parts as is mechanically possible. But since 40 bar of helium would be trying to slip through this gap, it would never be a perfect seal, and the power lost would be from the most important place, the place where the piston was forced back by hot helium. And with both conceptions—the aluminum rings on the magnet paddle, and the skirt on the piston—there would be leakage losses because of the lack of absolute roundness: nothing fabricated in a machine shop is *perfectly* round, only very round. The team members assumed they could make the parts round enough.

When the 5K team went over this in a design review—a devil's advocate presentation to engineers with various expertise—William asked whether the parts were in the shop. To William, the ideas looked fine: clearance seals had worked—well enough—on other Sunpower machines, and the gas bearings were working in the coolers, so he wanted his people to get cracking. For every issue that Neill brought up—feed tubes for the gas bearings, attaching the new piston to the old magnet paddle, and so on—William would draw a spidery line on the drawing for possible solutions. On the drawing, it looked simple, and simply and easily done.

Ideas are easy, hardware is hard. It took the 5K group most of a week to figure out a satisfactory plan for attaching the new piston to the magnet paddle, and even this plan wasn't particularly satisfactory. It involved drilling a ring of bolt holes in the very edge of the magnet paddle itself, which involved clamping the paddle in the milling machine, which involved accounting for its lack of concentricity, which involved trying to mirror its lack of concentricity on the piston skirt itself, which involved. . . . William's offhand comment in the design review about the leakage issue for the magnet paddle gas bearings—"the gas will make the magnet paddle blow itself round"—became a sort of mantra as the team struggled to replace, virtually wholesale, about a quarter of the most important parts in the machine. When the aluminum gas bearing rings came back from being anodized (a process that gives aluminum a very hard coating of aluminum oxide), one ring was impressively round; the other had somewhere along the line been dropped: it was the least round object anyone can remember having seen since Bob Black left some plastic parts in the curing oven overnight. It was a Salvador Dali–like round ring. Todd Cale was there when the parts were unpacked. "Shit," he said, "that'll blow itself round."

As they came closer and closer to having working hardware again, the niggling over money with Cummins was blowing itself round as well. The DOE Mission III contract was signed, and Cummins would be thrilled for Sunpower to start billing on the new contract—just as soon as it hit the milestones from the *previous* contract. (There were periods of time when Abhijit and Neill had the milestones firmly in mind—but different milestones. At one point, Neill suggested they simply tell Cummins they had hit all the milestones, and count on Cummins's not knowing what they were supposed to be doing, either.) As soon as the team ran the machine at six kilowatts for forty hours, Sunpower could start billing on the Mission III contract. This was both an incentive and an irritation: Cummins now had the money—or the contract for the money, but Cummins was playing hardball. William lobbied unsuccessfully for some of the 5K work to be billed to Mission III—the work that was ultimate-machine oriented, but Cummins wasn't interested. So there were, in essence, two incentives for the six-kilowatt, forty-hour run: one was the cash incentive from the previous contract, the other was a million and a half dollars in billable work. To a company that had just skipped a payday, this was a mighty incentive. Out of money was putting it nicely: Sunpower was not only out of cash to the point that it wasn't able to pay its employees; it was technically in default on its line of credit with the local bank. (Cucumber money had been earmarked for paying down the line of credit once a year, as the line of credit specified; no Cucumber, no pay down, unhappy banker.) "So, you guys need any help?" became the way that people began greeting the 5K team, a far cry from asking when it would start pulling its own weight. Team members tried hard, in 5K style, to do the calculations, do the drawings, hit their marks, but there was suddenly something to be said for being ignored; one didn't then have a constant string of coworkers hanging around to fetch tools.

As the new parts began to take shape, the rest of the design work went on—the fussy things, the things that no one really had any patience for but that had to be done. Neill would spend hours sitting next to Todd Cale in the drafting room, closing his eyes, envisioning, and giving Todd a dimension or a coordinate. It didn't look like heroic work; no shouting, no banging on something with a hammer, no beads of sweat. When the gas bearings were tested on the magnet paddle, it was almost a metaphor for this stage of their progress: one of the qualities of the gas bearings is that while two metal parts are ostensibly rubbing against each other (one is actually riding on a cushion of air), there is no sound: the magnet paddle floats, silently, while everyone around wants to hear the noise of productive hard work.

The paths for the tiny metal tubes that would carry high-pressure helium to the gas bearings was painstakingly worked out, as was the check valve arrangement that would pump the bearings up. The last conceptual issue, still unresolved as the hardware came together, was how to center the piston: centerports and clearance seals are two ways of saying the same thing: both provide a leakage path from an area of high pressure to one of lower pressure. But a clearance seal is not a controlled leak: it leaks a little bit all the time. So centerports were out, and an "in limiter" was in—another check valve arrangement, this one supposedly opening up when the piston reached a certain point in its cylinder. On paper, the team's fussiness made the new design look neat and well thought out; in the lab and the machine shop, there were parts everywhere, and time was passing. The first run of the "new" 5K was accomplished without the gas bearings' working at all, although the magnet paddle did rest on the anodized aluminum rings. (A "wear band" of Teflon-like material was glued inside the magnet paddle to provide a running surface.) Members of the team knew they didn't have it all worked out, but they wanted to see how close they were—if they were in the right church, wrong pew, or if they were in a different place entirely.

They surprised even themselves. They were back up to 60 Hz, for one thing, and thus gained a modicum of power just from that. But they hadn't expected to get *so much* more power: 5,500 watts without even stroking the machine out, five and a half kilowatts with no centering scheme and no gas bearings. They felt like clever chaps indeed.

Beej then set a date for a six-kilowatt run and planned for John Bean and Rocky Kubo to come from Cummins to see them hit their marks. In the interim, they assembled, ran, and disassembled the engine eleven times in twenty-three days, debugging the gas bearings, debugging the in-limiter, tweaking the machine: the phase angle would be off, and they would make adjustments; then the piston would run in, and they would make adjustments; then the phase angle would be off and the piston would run in, and they would make still more adjustments. They were watching the gas bearing rings very closely, since the magnet paddle had not, unfortunately, blown itself round: it had very little friction, but it was not perfect, especially not the ring on the "in" end of the paddle; no matter what they tried, it rubbed. They ended up taking it out completely, the magnet paddle supported only by the piston skirt and the remaining gas bearing ring. One whole week was spent on a maddening development in the in-limiter: the in-limiter was designed to be adjustable, with a threaded insert that could be turned in or out depending on where the piston was running, a sort of adjustable centerport. In one series of runs, the piston ran in, and the in-limiter was adjusted; the piston still ran in, though, and the in-limiter was adjusted some more. Then the piston ran *way* in. Shade, who had been adjusting the limiter, was given a pop quiz. Did he know left from right, in from out? "Raise your right hand *quick*." Another adjustment, and the piston ran way, way in. Neill Lane didn't sleep a wink that night: it made *no* sense—it was like, he said, "letting air out of your automobile tire and having it blow *up*."

It turned out to be a classic dynamic anomaly: the in-limiter, although they didn't know it, had been operating at the very fringes of its range when they started adjusting it. Beyond a certain point, they knew, it would do nothing—they just hadn't known they were at that point to begin with. Shade rebored the in-limiter, and the piston ran out satisfactorily.

Four days before John Bean and Rocky Kubo were due, they ran the machine and hit six kilowatts for a time; as the engine ran over the course of an hour, the power degraded, as power is wont to do—the machine gets hot, the helium gets hot, the bearings expand. They were close, but not quite there: they wanted an hour run at six kilowatts, and a day later, with some minor adjustments, they had it—one hour, steady state, 6,100 watts. They knew they were running at the upper limits of the machine—a temperature of 670 degrees Celsius, for example, and a pressure of 40 bar—but they had the power, and they had it when they needed it, and they could generate it at will. William was summoned, and he stood and watched the engine run for almost the full hour, watched the power, watched the efficiency numbers level out, watched the green Lissajous plot on the oscilloscope. "Well," he said about halfway through, "this has worked out quite nicely, hasn't it?" And then, a little later, "Good job, there, uh, Neill, and Beej and, uh, Dave."

And William immediately got on the phone to Cummins and moved the goal posts one last time. Don't bother coming, because it's a done deal—the machine, he was overheard saying, is working "absolutely perfectly." The forty-hour run would take Sunpower and its (unpaid) employees a solid week to accomplish, a week they could more *profitably* spend billing on Mission III. "At this point, you're really proving nothing; we know what we need to know," William said, and a compromise was reached. Eight hours, steady state, six kilowatts, and they could start billing on Mission III.

So, two days later, Neill and Abhijit and Dave Shade fired up the engine at a little after five in the afternoon; by five-thirty, they were at 6,100 watts, and they settled in, watching the engine run, watching the numbers on the data screen, watching the phase plot, listening to the hum of the engine. Neill insisted on written data as a backup—they weren't going to do this twice. So every ten minutes between five-thirty and two-fifteen the next morning, someone would read the numbers off the data screen and someone else would write them down: the time, the pressure, the heat pipe temperature, the frequency, the power, the piston amplitude, the displacer amplitude, the temperatures of the regenerator, the back end, the compression space, the voltage, and the current. At two-thirty in the morning, when there was nothing left to prove, Neill Lane went into the darkened offices and plotted out a chart showing power over time. But for one blip when they thought they heard a funny noise and they cut the power back, the dots creep across the time scale on the line marked six. The last thing Neill did that night was photocopy this graph and write across it, "Six kilowatts, eight hours! Glad we could be of help!" He taped it to Sunpower's front door and went home. He took the next day—a day conceivably billable to Mission III—off.

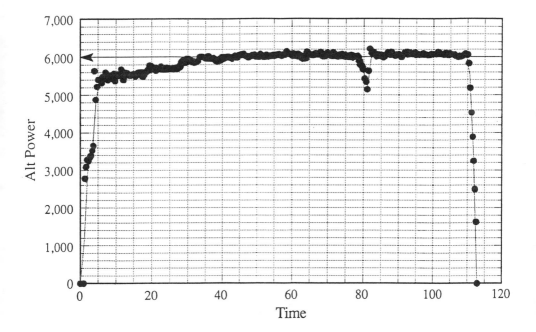

Chapter 13

The Carnival Comes to Sunpower

B Y THE TIME Elaine Mather fields the umpteenth call
about the noise, she has a headache herself. Still, she
scrounges a piece of poster board and, with a marking
pen, writes on it, "We apologize for the inconvenience/Engine testing
in progress." As an afterthought, she dates the sign and adds, "Today
only," as though Sunpower were having a sale, a sale on noise. She
carries it out to the Sunpower gate on Byard Street and wedges it into
the rusty Cyclone fence.

The half dozen neighbors who have been watching Sunpower
with undisguised annoyance for the past two and a half hours come
close enough to read her sign, and then go back to muttering to themselves. William dispatched Hans to the lumberyard for some large
sheets of plywood to strap up around the generator in an attempt to be
neighborly, but it was mostly cosmetic. The noise is, indeed, horrendous.

The morning noise was from the world's least-impressive forklift—borrowed from a friend of a friend of Sunpower and looking as
though it had been abandoned for decades—as the Sunpower machinist Ray Klinebriel, who had operated a forklift in Vietnam, wiggled a
massive air compressor out of the back of a Ryder rented truck. The
forklift's unmuffled exhaust was loud enough; when the 200-kilowatt
diesel generator was fired up, it made the forklift sound like a cat's purr.
A 200-kilowatt diesel generator is seldom run in any sort of residential
area, because 200 kilowatts is enough electricity for a rather large traveling carnival.

This generator in particular (rented by Cummins Power Generation, Inc., from Cummins Mobile Power, Inc., both subsidiaries of the Cummins Engine Company, Inc.) is noisy even as 200-kilowatt diesels go; it was manufactured in Minnesota by the Onan Corporation (another subsidiary of Cummins Engine) when Sunpower was still the glimmer of an idea in William the Beale's eye. But the price was right, and Sunpower needed the juice.

The need for 200 kilowatts of electricity is twofold; part of it goes to run the air compressor; the rest, to run a small rack of quartz lamps, the sort of lamps one would use to light up an auto racing track or a very large prison yard. The lights are concentrated, though, behind a square pane of glass about two feet square, and focused in the delicate hemispherical dome of a new Thermacore heat pipe. Rather than being run by electric heaters like the test cell heat pipe, this heat pipe is the real thing, the one headed for Abilene, Texas, and a date with a solar concentrator. The light bulbs burn so hot they must be cooled with compressed air—hence the air compressor, which is louder than the generator. The bulbs are hot because they are providing 30 kilowatts of energy to the concave receiver of the heat pipe, acting as a portable—and noisy—sun.

Attached to the hemispherical heat pipe is the 6K—rechristened in the Mission III project to reflect its new design power, but its guts are the same 5K that Sunpower undertook to design and fabricate for Cummins once upon a time. Outside the test cell stands a large water tank with an overhead spout connected to a large electric motor and water pump. The last milestone that Sunpower will hit on what is now officially referred to as "Engine/Alternator 1" is pumping six kilowatts of water on sun. The diesel-powered sun presides at the last dress rehearsal.

The yard around Sunpower does resemble a carnival of sorts—cables and wires strung everywhere, rental trucks, lots of people milling about. As the heat pipe heats, Neill Lane does a shirt-cuff calculation of true efficiency: 300 kilowatts of diesel fuel, 200 kilowatts of electricity, 30 kilowatts of light, 6.5 kilowatts of electricity, 6 kilowatts of water. Put this way, what they are doing looks ludicrous. But if things go well, by next week the numbers will have changed, because the diesel generator and the quartz lamps will have been replaced by the nonportable sun, and the 6 kilowatts of power will be pure gravy, the sun doing useful work, by means of a Stirling engine and courtesy of Sunpower, Inc. Noise? What noise?

As the rather scruffy group of technicians and specialists debug the setup (the Thermacore guys were on the road all night) and prepare for the run, Neill Lane wanders back inside to the lab, where even a first-time visitor would notice what seem to be positively *heaps* of shiny stuff all over, spilling out of boxes, covering bench tops, stacked on tables. It is Christmas in April, or something like it, at Sunpower. All around Hans's workbench, in multiples of three, parts for the first batch of solar 6K engines have begun to arrive. Gone are the days when Hans would get Neill's goat by asking him what he wanted done next on the "*three* kilowatt engine."

The design has been validated in the most concrete way possible: plans for engines beyond the prototype. There are three pressure vessels, painted primer red and emblazoned with official-looking test certification plates; three transition pieces, three sets of bearings, three piston cylinders, three displacer rods, three displacer domes, three displacer canisters, right down to three sets of bearing pads and three "salmon plates," which have nothing to do with fish, but are among the few parts of the engine that have remained a part of the engine since day one of the axisymmetric design process—"salmon" refers to the color by which the part was drawn in the CAD program. (As such, the salmon plate is a sort of sentimental favorite part in the engine, the one that everyone always asks about.)

One year earlier, even the most optimistic person at Sunpower would have bet 1,000 to 1 against such an array of hardware, because the project was expected at that time to be moribund in days. Now they are about to make power on-sun—one coup—and are surrounded by parts for three more engines, a second coup.

For months, Neill and Beej and Todd Cale, the draftsman, methodically made their way through every single drawing of the design, checking part numbers, checking dimensions, checking specifications, and shipping the drawings in batches, via computer disk, to Cummins. In a mirror of the inches versus millimeters language barrier in the machine shop, it turned out that Sunpower uses for its drafting one of the two most widely accepted computer programs in the world. Cummins uses the other one, though, so the drawings had to be "translated" by the computer before they could be sent off. But translated they were, and Sunpower sent a complete set of drawings off to the sponsor—as a real company might do.

Beej and Neill pick the parts up, put them down, run their hands over the surfaces, reach into shipping crates to pull out yet another piece of the 5K puzzle. The Cummins big boys are present—John Bean, Paul Richart, John Christensen—and they are looking a little smug. They have a right to be. The parts drawings had been jobbed out to a small machine shop in Seymour, Indiana, a shop that had matter-of-factly cranked out the parts—from Sunpower's drawings. (Cummins's modifications to the drawings consisted of removing the Sunpower name block from the drawings and substituting its own.) The parts look perfect. Cummins has a flashy "coordinate measuring machine," which electronically measures a part and matches the measurements to the part drawing. The technicians' notations are on each drawing, and the parts are correct to the third decimal place, which is as close to perfect as one could ask.

Jarlath McEntee comes through and fondles the parts as well, looking more relieved than he has in quite a while; a few weeks earlier, he was feeling put upon in the extreme, working under extremes of pressure on design work for three different projects, glued to the ALGOR computer system, and grousing that he was doing the work of three people for half pay. This week, he is back on full pay, something that a year ago would have been a terrible bet to make as well, given that the DOE heat pump was unfunded and the only big heat engine in the place that was funded at all was scraping by on pennies. Six months before, one would have been hard pressed to assemble three engineers at Sunpower on full pay. Now, as in the old days, they seem to be everywhere.

Neill has work to do, but he can't resist playing around with the 5K parts a little more, sliding the bearings into their sleeves, twisting a displacer dome onto a canister and marveling at the threads, picking at an imaginary burr in an O-ring groove. He looks pleased, extremely pleased.

Inside, though, he is deeply concerned, as worried as he has ever been about the 5K—the 6K, he has to keep correcting himself. He and Jarlath have perhaps identified the showstopper to end all showstoppers on the engine, a problem that no one had ever previously identified, and one for which there is at this moment no apparent solution that doesn't sound like a terrible—indeed, deeply ironic—idea. Neill believes he has figured out what at first seemed to be a curiosity—the Mystery of the Growing Heater Head. The news is bad. In the hall, he shakes his head. "I *hate* always being the damned bringer of bad news," he says. "You get *no* credit whatsoever for a truckful of parts, but when you identify a problem, everyone gets *pissed*. Sometimes I *hate* this place." He sneaks outside for a cigarette and thinks about what he has just learned.

The Mystery of the Growing Heater Head has never been mentioned in all the literature about heater heads and heat pipes, although

Neill is certain that it is an endemic problem: no one has ever noticed it before, or at least figured it out. The original heater head for the 5K gradually changed shape over time, after tens of cycles of heating up, running, and then cooling down. The change wasn't appreciated until the bolts on the head began to break in two, and the first thought—that the stresses on the bolts had been underestimated—turned out to be wrong: twice as many bolts, meticulously torqued to a precise tightness, hadn't helped. (When you double the number of bolts holding something together and they *still* break, it is axiomatic that all the bolts in the universe won't solve the problem.) So Neill and Jarlath, with the help of ALGOR, a computer program, had designed a different head, one that was beefy where the previous design seemed to be weakest, and they had been rewarded by an almost immediate failure. "Stress-relieving" the head hadn't helped either; after each heating/running/cooling-down cycle, the head was measurably stressed, growing in length along the order of a tenth of a millimeter, which sounds tiny; but because the growth was cumulative, it meant that each time the engine ran, the head changed shape by about a tenth of a millimeter. Over ten cycles, the distortion would be a millimeter; over a hundred cycles, it would be ten millimeters, if the head held together that long. Which of course it would not: even one millimeter of change was far beyond what was considered reasonable, acceptable, or safe, because the change in shape of the heater head came at a tremendous cost in strength of the head itself. Long before it would grow a millimeter, it would yield.

In the best case, one of the two hundred heater tubes would begin to leak, performance would degrade, and the sodium in the heat pipe would be contaminated by helium. In the worst case, the dome of the heater head would yield catastrophically, and the damn thing would blow up.

As this slowly became apparent, it posed a practical problem for Sunpower; it needed to run the engine in order to debug the water pump rig, but running the engine made the heater head weaken. Dave Shade noted acidly that Abhijit's solution had been to declare the test cell off-limits to all nonessential personnel for reasons of safety, which meant that it was just Dave Shade and a potential bomb. "I guess I'm nonessential," he says. "It would make more sense to get all the nonessential personnel in here, but what do I know?"

Shade's gallows humor notwithstanding (he is always trying to gather the "nonessential" personnel in one place; the whole notion of someone being nonessential amuses him—why are they still being paid?

he wonders), the problem is serious. It had long been taken as fact as Sunpower (and, apparently, in the rarefied world of heater head design everywhere else) that stresses and strains are calculated at steady-state operating temperatures and pressures: a piece of high-strength alloy steel sitting on a bench at atmospheric pressure and room temperature does not provide a boundary condition of any interest whatsoever. It would of course just sit there. So all the calculations for the head design were made at operating pressure (40 bar) and operating temperature (650 degrees Celsius). It has always been presumed that at this extreme the stresses in the head would be at their maximum, and so a head was designed to contain these stresses. But the 5K head was still yielding, at the rate of about a tenth of a millimeter per cycle, during operation. Why?

The answer came from an examination of data collected while the heater head was being brought up to temperature. The company that made the heat pipe, Thermacore, had provided amusingly brief instructions for operation—a half dozen typed pages that, boiled down, said to turn the heat pipe on at a low temperature for a little while, crank it up about halfway for a little while longer, and then turn it up the rest of the way. ("Safe operation of a sodium heat pipe is predicated on a basic understanding of heat pipes combined with common sense," the instructions begin. Hans Zwahlen once pointed out that more operating instructions came with his VCR.) The logic behind a warm-up period was simply that the sodium should be softened up, then slowly vaporized, before one goes to operating temperature, so that no hot spot in the heat pipe would burn a hole in the casing or burn out one of the electric heating cartridges, much the way one might warm up the car for a few minutes on a cold morning or thaw a chicken breast in the microwave. But the temperature readings collected during the warm-up phase almost made Neill's hair stand on end: the assumption was always that the heater head heated up gradually and evenly during the warm-up phase—there had never been any suggestion to the contrary.

But what Neill found when he plotted temperatures on the heater head over the time of the warm-up period were *huge* differences in temperature between places that were only a few millimeters apart. Rather than the tens of degrees difference that he—and everyone—thought there would be as the heat pipe approached isothermal conditions, there were differences of as much as *four hundred* degrees Celsius between, say, thermocouple 5 and thermocouple 6, which were about a finger's width away from each other. As any junior engineer would know, such gigantic differences in temperature were a perfect recipe for generating huge internal stresses. As a child in Rhodesia, Neill Lane and his friends liked to put marbles in the freezer overnight, and then drop them into a pot of boiling water. The desired result was a beautifully "crazed" marble, with a network of thousands of cracks. Sometimes, though, the marble instantly disintegrated into a sandlike swirl of grit at the bottom of the pot. It was these marbles that Neill thought of when he saw the temperature differences in the heater head during warm-up.

The obvious solution suggested gallows humor worthy of Dave Shade: simply heat the engine up very slowly—preheat it, in other words, over a period of hours, rather than minutes. In a sort of shirt-cuff calculation, Neill figured that a warm-up time of about three hours would work, which meant that in the ultimate incarnation of the 5K, when the engine was on sun (and even if the mirrors could be configured in such a way as to permit such a slow warm-up), instead of getting useful electricity at, say, eight-thirty in the morning, one would get no useful electricity until about noon. In this sort of arrangement, the cost of a kilowatt of electricity from the 5K would double. In an industry where kilowatt-hours are priced out to hundredths of a cent, the doubling of the price would make you laugh, if it didn't make you cry. To make this work, the warm-up period would just have to start before dawn.

The other obvious solution was worse; since the problem was related to the heating-up and cooling-down cycle, keeping the heat pipe warm when the engine wasn't operating would be a quick fix. "Ehhh, make five kilowatts for eight hours, and use two and a half kilowatts for sixteen hours keeping the heat pipe warm," said Neill ruefully. "That's perpetual motion: 'No need to worry about tying into the electrical grid, guys; we won't need to; we're going to use all the power we make keeping the engine warm.' " This was either funnier, or sadder, depending on one's capacity for irony, and depending on how much time or money one had invested in the project.

Although a cruel god might have chosen a worse moment to conjure up this problem, it's hard to imagine it. Neill walked back into the shop where all the parts for the first, fledgling production run of the 5K were stacked: three pressure vessels and so on, right down to little plastic bags of bearing pads. What had been a rush of accomplishment and pride and even a little exultation was somewhat overshadowed by the heater head. "Ehh, they're going to want something to put on the front of these, aren't they?" Neill says. He runs a hand over a salmon plate, thoughtfully. "We just don't know what it is right now."

In the short term—the short short term—the interim solution is simply to warm the head up slowly. The arrival of the on-sun heat pipe took some of the danger away, in that it would be, at the very least, perfectly safe for dozens of cycles; a longer-term solution was still mysterious.

Outside, the carnival gang is getting closer; after a classic Sunpower conversation over how to ground—or not to ground—the diesel generator (two electrical engineers, three mechanical engineers, two electricians, five kibitzers, two technicians, and William), the quartz lamps are fired up and the process of "coming on sun" is initiated. With the diesel generator now pulling a real load, it gets louder. Then it blows a circuit breaker. Another conference ensues, the lights come on again, the breaker blows. It turns out that while it might be a carnival-sized generator, it came equipped with a toy train–sized circuit breaker. The solution is to run the air compressor off Sunpower's own electrical supply and to leave the generator just for the lights. The connections are made, and the lights again begin to warm up the heat pipe.

What makes this all so absorbing to the 5K team is simple: the plan is to run the engine on the lamps, pump six kilowatts of water for an hour, take the whole rig apart, and *ship the engine* to Abilene, Texas, in a Ryder rental truck, with John Bean of Cummins at the wheel. If things go well, when John pulls out of the driveway and turns right onto Byard Street, the first machine will have left the door, on its way to Texas for running, on sun, while Sunpower fabricates and debugs three more engines, and Cummins prepares to fabricate three more engines itself. There are problems, and there will be more—with the concentrator, with the circulating pump for the cooler, with the load controller—but they will not be *Sunpower's* problems. Sunpower's problems end when John Bean leaves. For a day at least.

As the heat pipe comes up to temperature, Neill Lane sits on the tailgate of the Ryder truck and idly imagines following the truck until John Bean stops for the night, and then *stealing* it, but he decides he doesn't really want an engine after all. Perhaps swap the truck with an identical one full of furniture? "Wouldn't you like to see John Bean's face when he *raises* the door and an overstuffed chair tumbles out!"

Despite the heater head problem, Neill can't help being a bit boisterous: being outside helps, and being near to finishing a stage of the project that has absorbed him for two years. Then too, Neill is fresh from a long meeting in which it seemed that Cummins was getting really serious about the engine: a vice president from Cummins Engine (rather than the subsidiary Cummins Power Generation), surrounded by his technical mavens, had listened to Neill go through the 6K design step by step by step. Neill talked for five hours, virtually nonstop; at the end of the meeting, the VP said that the solar Stirling program didn't sound like an *R & D* program; it sounded like a commercial program, a real product. "Is this program starved for resources?" he asked at one point. "If we think about this as a commercial product, it sounds to me like we ought to commit more money." This is something that someone says to Sunpower rarely, if at all. "They want to give us more money!" was the optimistic shorthand summary of the meeting, to which William said, "Well, of course they would; we've got a very good design here." He made it sound matter-of-fact.

With the heat pipe up to temperature from the quartz lamps, there is nothing to do but run the engine, and Abhijit starts it up. But for the maddening two days when the feedback springing kept the engine from starting, one of the hallmarks of the 5K's "personality" has always been that when the starter is hit, the machine comes to life. It does so again, the hum of the alternator drowned out by the roar of the diesel generator, and then accompanied by the rush of water: outside the test cell, the pump is pumping water in an endless loop through the water tank. As Beej strokes the machine out, the generator gets louder under the increased load, the water gets louder as it is pumped faster, and there are, for a matter-of-fact outfit, a fair number of "oohs" and "ahhs" as the engine hits six kilowatts, then six and a quarter. William comes in, hands in his pockets, and room is made in the test cell for him; he squints at the data screen, wanders outside to watch the water pump pump water, comes back in. He makes an offhand comment or two to John Bean, peers in at the quartz lamp arrangement, and then says offhandedly to Neill Lane, "Very satisfactory, hm?"

To which Neill Lane replies, "Piece of *cake.*"

Epilogue

T HE EPA COMPRESSOR was a stunning and clear technical success, and John Crawford is in the process of "shopping" it to the usual suspects—the largest manufacturers of refrigerator compressors in the world. Most people at Sunpower still find it to be an inferior solution, but the director of global change was highly impressed: the prototype that Sunpower shipped for independent testing is something like 20 percent more efficient than a comparable CFC-based machine, in an industry where millions of dollars are spent to make machines tenths of a percent more thrifty. EPA Administrator Carol Browner has said that the Sunpower cooler "will do for the refrigeration industry what the microchip did for the computer"; the refrigeration industry, however, is no Silicon Valley. People in the industry are polite, but noncommittal. No one has at this time spent even a penny to investigate using the EPA compressor.

The Cucumber, while not exactly an orphan, is still something of a neglected stepchild. Cucumber, Inc., decided that while it wouldn't provide Sunpower with any more funding itself at present, it would steer one of its own favored customers to Sunpower. So Sunpower and Favored Customer, Inc., are negotiating. Cucumber, Inc., is still "evaluating the market potential" for the Cucumber, which by all accounts is huge. Modified Cucumbers—the "Super Cuke," as David Berchowitz calls it—have been demonstrated for a half dozen other companies, most of which are interested in using it for superconducting purposes. License agreements have been written, but no one is manufacturing them yet.

The CFC-free refrigerator prototype remains an orphan; it still works marvelously when demonstrated for a manifestly uninterested audience, but no one with whom Sunpower has spoken has considered doing anything with it until 2007 at the earliest.

The DOE heat pump is at present the subject of the longest report and the most extensive series of tests in the history of Sunpower, Inc., a place where a lot of long reports and extensive tests have been done. Present hopes are that a new administration in Washington will lead to funding for a next-generation machine.

The small engine was shipped as a prototype to a Swiss research consortium, which promptly began modifying it to the point where it no longer works. William, Al Schubert, and Chuck Howenstine made another prototype for John Grandinetti, who took it back to China for a further round of meetings. It is still "William's machine" and is still starved for funding.

The original 6K machine is running on sun in Abilene, Texas, at the Cummins test facility there. The second 6K was shipped for testing to the Pennsylvania Energy Office, and is running on sun (when there *is* sun) there. Four more 6Ks were finished and shipped before the focus shifted to "Phase II"—a nine-kilowatt, true forty-thousand-hour machine. The heater head problems were addressed by a new head design, a previously untried "torospherical"-shaped head that as of this writing is not growing much at all. The first 9K was scheduled to be shipped in August 1993.

In a meeting with Cummins Power Generation, Inc., Sunpower was told that CPG had not received DOE funding for the ultimate ultimate machine—a twenty-five-kilowatt engine that CPG, it turned out, had planned to pursue not with Sunpower at all but instead in conjunction with the "Clever Fellows Consortium," a Stirling-machine company that was not Sunpower but, rather, one of Sunpower's few competitors. This bit of news floored most of the Sunpower people, because they thought the work they had done for Cummins entitled them to something like most-favored-client status. William went into orbit at the news, talking darkly for weeks about the perfidy of spon-sored research. But it was a moot point without funding, and Cummins and CPG continue their sometimes uneasy alliance. At this writing, Cummins Engine has a new CEO and has not decided whether to make a full-scale commitment of its own money to solar Stirling, or whether to continue on the government-funded path of cost sharing. But Cum-mins does own the rights—and the hardware—for what is at this time the largest and most serious attempt at solar Stirling power generation in the world, and it is performing a cost study for a plan to manufacture one million machines a year. To run the United States on Stirling machines, Cummins will need somewhere around a hundred and ten million machines, at a shirt cuff–calculated cost that William estimates as something like the U.S. military budget for about ten years. They already have the first six.

Index